Paying the Price

The Status and Role of Insurance Against
Natural Disasters in the United States

Howard Kunreuther
Richard J. Roth, Sr.
Editors

JOSEPH HENRY PRESS
Washington, D.C. 1998

JOSEPH HENRY PRESS • 2101 Constitution Avenue, N.W. • Washington, D.C. 20418

The Joseph Henry Press, an imprint of the National Academy Press, was created with the goal of making books on science, technology, and health more widely available to professionals and the public. Joseph Henry was one of the founders of the National Academy of Sciences and a leader of early American science.

Library of Congress Cataloging-in-Publication Data

Paying the price : the status and role of insurance against natural
   disasters in the United States / Howard Kunreuther, Richard J. Roth,
   Sr., editors
      p.   cm. — (Natural hazards and disasters)
   A collection of 9 essays.
   Includes bibliographical references and index.
   ISBN 0-309-06361-2 (alk. paper)
   1. Insurance, Disaster—United States. I. Kunreuther, Howard.
II. Roth, Richard J. III. Series.
HG9979.3.P39   1998
368.1'22'00973—dc21   98-12369
                                      CIP

Printed in the United States of America.

*To Gilbert White, for his
foresight and leadership*

Foreword

THE NATION'S FIRST NATURAL hazards assessment got under way in 1972 at the Institute of Behavioral Science at the University of Colorado. Funded by the National Science Foundation and led by geographer Gilbert White and sociologist J. Eugene Haas, it was an interdisciplinary effort involving scores of policymakers, practitioners, and scholars from across the nation. Its purpose was to assess our knowledge about natural hazards and disasters, to identify major needed policy directions, and to inventory future research needs. The volume summarizing this endeavor, *Assessment of Research on Natural Hazards in the United States*, published in 1975, was a landmark in what was then a new field of study.

A quarter of a century later we find ourselves in a national conversation about how natural hazards mitigation can result in disaster-resilient communities. This conversation began in the early 1990s among a few individuals working in federal agencies and academia, and was articulated at a workshop in Estes Park, Colorado, in 1992, which was attended by many of the nation's leading natural hazards experts. They concluded that a second assessment of hazards in the United States was needed, and that its unifying theme should be sustainable development, or development that enhances the capacity of the planet to provide a high quality of life now and in the future. A subsequent workshop in 1994 brought many of the same experts together to discuss and formulate an agenda for this second hazards assessment.

This book, which ensues from the second natural hazards assessment, is one of a series of books published by the Joseph Henry Press. A select group of experts were invited to expand upon their necessarily condensed contributions to the second assessment by developing individual works on major themes in the natural hazards and disasters field, including land use management, risk assessment, disaster preparedness and response, and mapping. This volume on natural hazards and insurance was written by a group headed by Howard Kunreuther, of the Wharton School's Risk Management and Decision Processes Center.

Paying the Price examines the roles of private and public insurance in dealing with natural disasters, discussing insurability conditions for natural disasters; the changing demand for residential disaster insurance in the United States; the challenges insurers face in providing insurance from earthquakes, hurricanes, and floods; and the functions of state insurance regulators. The book describes varying levels of state and federal involvement in insurance programs as exemplified by the California Earthquake Authority, Florida Hurricane Catastrophe Fund, and the National Flood Insurance Program. The authors take the position that the economic costs of natural disasters are too high and likely to soar in the future unless steps are taken to change recent trends. They discuss the role that insurance and mitigation together can play in reducing future losses from natural disasters. The program they propose for reducing disaster losses and financing recovery is characterized by joint efforts between insurers and other stakeholders, and by the use of strategies that combine insurance with monetary incentives, fines, tax credits, well-enforced building codes, and land use regulations.

The second national natural hazards assessment was funded by the National Science Foundation under Grant Number 93-12647, with supporting contributions from the Federal Emergency Management Agency, the U.S. Environmental Protection Agency, the U.S. Forest Service, and the U.S. Geological Survey. The support of these agencies is greatly appreciated. However, only the authors are responsible for the information, analyses, and recommendations contained in this book. Very special thanks are extended to J. Eleonora Sabadell and William A. Anderson of the National Science Foundation for placing their confidence in us to carry out this mission.

DENNIS S. MILETI, *Director*
Natural Hazards Research and
Applications Information Center
University of Colorado at Boulder

Preface

THIS BOOK IS ABOUT THE ROLE that insurance, in combination with other policy tools, can play in reducing losses from natural disasters and in providing financial protection for those who suffer property damage from disasters. When we began this study four years ago, the insurance and reinsurance industries were reeling from the catastrophic losses many firms had experienced from Hurricane Andrew in Florida (August 1992) and the Northridge earthquake in California (January 1994). Many in the industry had grave doubts about whether insurers and reinsurers could continue to provide protection against wind damage from hurricanes and shake losses from earthquakes. At around the same time, other positive developments promised to aid insurance firms in offering coverage against these risks. Science and engineering research studies were yielding more accurate information on the probability and consequences of disasters of different magnitudes. New advances in information technology enabled large databases on population and property at risk to be manipulated more easily and rapidly than in the past. Insurers and reinsurers were beginning to employ catastrophe modeling when determining the type of risks to include in their portfolios.

It is against this backdrop that this study was initiated as part of Dennis Mileti's broad-based second assessment of natural hazards. Our purpose was to bring together the country's leading experts on insurance and reinsurance to examine how insurance and reinsurance have provided protection against natural disasters in the past, and how they can play a more creative role in the future. There was general agreement among these experts that insurance can form the cornerstone of a disaster management program. However, there was also consensus that, to be successful in this regard, insurance has to be closely tied to other policy tools such as incentives, fines, tax credits, well-enforced building codes, and land use regulations. Furthermore, the insurance and reinsurance industry needs to work closely with other parties concerned with natural hazards, notably banks and financial institutions, the real estate and construction industries, and public sector agencies at the local, state, and federal levels.

The first three chapters of the book provide a general overview of the problem of dealing with low-probability, high-consequence events, the nature of the supply of insurance, and the demand for insurance against natural hazards. The middle portion of the book focuses on three case studies: the earthquake risk in California, the hurricane risk in Florida, and the changes in the National Flood Insurance Program over time. The final portion of the book focuses on strategies and institutional arrangements for reducing losses from natural disasters: the role of hazard mitigation coupled with insurance, the nature of insurance regulation, and a proposed program for reducing losses from natural disasters.

The principal focus of this book is on the impact that natural hazards have on the residential sector. The insurance crises in Florida, California, and other states were triggered by public concern that many insurers would discontinue providing homeowners with coverage against natural hazards. We recognize the importance of the commercial sector, and include a review of the availability of insurance for commercial property as Appendix A. At various points in the book, the authors note the great potential of catastrophe risk models in evaluating potential losses from natural hazards; a perspective on the challenges and opportunities of these models is provided in Appendix B.

This book examines a number of new developments in the insurance industry. The formation of public-private partnerships, as epitomized by the California Earthquake Authority at the end of 1996, is one example.

Continued change is inevitable. To help readers keep abreast of change in this field, we list a set of World Wide Web links to topics covered in the text, and describe in detail in Appendix C some key sites currently on the Web.

Many individuals have contributed to this book in a number of different ways. Those providing relevant material that was particularly useful to the principal authors are noted in footnotes at the beginning of each chapter. After a first draft of the book was produced, Richard Roth, Sr., worked closely with Robert DeChant and Don Seagraves in editing the material for style and consistency. DeChant was also responsible for the glossary of insurance-related terms.

At the Hazards Research and Applications Workshop in Boulder, Colorado, in July 1996, an early draft of the book was critiqued by a panel consisting of Jack Nicholson, Ellen Seidman, French Wetmore, and Eldon Ziegler, all of whom provided very helpful comments. In addition, we received helpful suggestions on drafts of different chapters from Jim Ament, Hal Cochrane, Clem Dwyer, Baruch Fischhoff, Jerry Foster, Rachelle Holland, Mark Johnson, Bill Martin, Mary Fran Meyers, Elliott Mittler, Tad Montress, Sean Mooney, John Mulady, Frank Nutter, George Segelken, Craig Taylor, Jeff Warren, Gilbert F. White, and Berry Williams.

In the course of writing this book, the authors had many helpful discussions on the issues associated with insurance and natural hazards with colleagues in both the academic community and the insurance world. Special thanks go to Ken Arrow, Ray Burby, Karen Clark, Graham Cook, David Cummins, Neil Doherty, Weimin Dong, Ron Eguchi, Ken Froot, Steve Goldberg, Robert Irvan, Gil Jamieson, Dan Kahneman, Paul Kleindorfer, Dennis Kuzak, Larry Larson, Scott Lawson, Chris Lewis, Peter May, Richard Moore, Ayse Onculer, Mark Pauly, Tony Santomero, Tom Schelling, Cliff Smith, Tom Tobin, Susan Tubbesing, Randy Updike, Susan Wachter, Sid Winter, Richard Zeckhauser, and Art Zeizel.

The production of this book was a collaborative effort on many fronts. Stephen Mautner, executive editor of the Joseph Henry Press, was extremely open to new ideas in producing this book and encouraged us from the outset to be creative in the use of graphs, photos, and maps to illustrate key points. Sharon Bolton, Walter Hays, and Ed Pasterick helped us to find disaster photos; Scott Lawson, Ed Pasterick, and Bob Klein were responsible for the maps included in the book. On the editorial side, Ann Perch and Marcia McCann worked closely with us in copyediting the book; we are very grateful to them for spending weekends and late hours to help us make our deadlines while remaining cheerful in the process. We are also indebted to Anne Stamer, who, with efficiency and good humor, coordinated the entire effort with the many coauthors throughout the four years of this project.

A special note of appreciation to Dennis Mileti, who made this whole project possible: his vision of the second natural hazards assessment, plus his tireless efforts in assisting each of the many subprojects under this banner, enabled us to move forward on our individual efforts. He also graciously provided us with financial support for meetings and editorial assistance through his National Science Foundation (NSF) Grant. We also want to thank William Anderson for his encouragement and support of this effort. Partial funding from an NSF grant (# 524603) to the University of Pennyslvania and from the Wharton Risk Management and Decision Processes Center at the University of Pennsylvania is gratefully acknowledged.

Finally we want to thank our wives, Gail and Liz, respectively, for being active commentators on the project since its inception. They helped us to better understand the challenges in communicating information on a topic that is not part of normal conversation at the dinner table or at social gatherings. We hope the excitement that our entire group feels about new ways of dealing with natural hazards is communicated in this book, even though we have no illusions about radically changing the world overnight. We hope that at least our country will be moving in the right direction. Otherwise, we will be Paying the Price.

HOWARD KUNREUTHER
RICHARD ROTH, SR.
June 1998

Contributors

HOWARD KUNREUTHER, *co-editor,* is the Cecilia
Yen Koo Professor of Decision Sciences and Public
Policy at the Wharton School, University of Pennsylvania, as well as Co-Director of the Wharton Risk
Management and Decision Processes Center. He has a
long-standing interest in improving how society manages technological and natural disasters, and has published extensively on the topic. Kunreuther is currently
a member of the National Research Council Board on
Natural Disasters and chairs the H. John Heinz III
Center Panel on Risk, Vulnerability, and True Costs
of Coastal Hazards. He is a recipient of the Elizur
Wright Award for the publication that makes the most
significant contribution to the literature of insurance,
and is a Distinguished Fellow of the Society for Risk
Analysis.

RICHARD ROTH, SR., *co-editor,* is a retired insurance executive and property/casualty actuary. He is a
Fellow of the Casualty Actuarial Society and a professional member of the American Meteorological Society, the American Geophysical Union, and the American Statistical Association. He has experience with

many aspects of natural hazards insurance and was Chairman of the Federal Insurance Administration's Write Your Own (WYO) Standards Committee for the National Flood Insurance Program. He is currently engaged as an actuarial consultant for the H. John Heinz III Center for Science, Economics, and the Environment for its Evaluation of Erosion Hazards project and is a member of the National Research Council's Board on Natural Disasters.

Contributing Authors

Jim Davis, California Division of Mines and Geology
Karen Gahagan, Institute for Business and Home Safety
Robert Klein, Georgia State University
Howard Kunreuther, University of Pennsylvania
Eugene Lecomte, Institute for Business and Home Safety
Charles Nyce, University of Hartford
Risa Palm, University of North Carolina
Edward Pasterick, Federal Emergency Management Agency
William Petak, University of Southern California
Richard Roth, Jr., California Department of Insurance
Craig Taylor, Natural Hazards Management
Eric VanMarcke, Princeton University

Contents

Paying the Price

CHAPTER ONE

Introduction

HOWARD KUNREUTHER[1]

T HE FINANCIAL COSTS OF natural disasters are on the societal radar screen. In the past they received relatively little notice except in the aftermath of a catastrophic event. The reason for our increasing awareness of the costs of natural catastrophes was succinctly summarized by two U.S. congressmen in an article in the *Washington Post* in 1995 entitled "Natural Disasters: A Budget Time Bomb": "Over the past five years the cost of natural disasters has been rising at an alarming rate. In that time, 11 catastrophes have cost the nation more than $1 billion each. Hurricane Andrew and California's Northridge earthquake together cost more ($28 billion) than what the government spends annually on running the federal court system, aiding higher education and pollution control, combined" (Emerson and Stevens, 1995).

Even in a year when there are no significant single events, the trend is continuing, as indicated by the following quote in the Munich Reinsurance Company's summary of disaster losses in 1996: "1996 was a normal year for catastrophes, with losses of over

---

[1]Robert Klein provided contributions to this chapter.

Houses in subdivision destroyed by Hurricane Andrew, 1992 (FEMA).

US$ 60 billion, of which US$ 9 billion were insured, and nearly 12,000 fatalities. However, the general trend toward ever-increasing numbers of catastrophes with ever-increasing costs is continuing" (Munich Re, 1997, p. 16).

As we will indicate in this book, in addition to an increase in the number of natural disasters, the dramatic rise in disaster losses has been caused primarily by a large increase in the population of hazard-prone areas, as well as a rise in the costs of construction. This book investigates the appropriate role of insurance to help stem this tide.

Insurance has an advantage over all other methods in the policy analyst's tool kit in that it rewards individuals prior to a disaster for investing in loss reduction measures through lower premiums, as well as paying these same people for damages suffered from a disaster. For insurance to be effective in both these roles, those who are at risk must bear a substantial portion of the costs of residing in hazard-prone areas. Otherwise they will have limited economic incentive to take protective actions and will rely on others to bail them out after the next disaster.

Insurance can play a key role in addressing two broad questions that are central to designing a hazard management program whose principal objective is to reduce future disaster losses:

- Who should bear the costs of making hazard-prone communities safer?
- Who should pay for the losses caused by disasters?

There are two criteria normally utilized in addressing these questions: *efficiency* and *equity*. By *efficiency* we mean the allocation of economic resources to maximize social welfare. Social welfare is defined by the citizenry and thus may vary from one political entity to another. A society that believes that every citizen has the responsibility to pay for the disaster losses of victims may find that taxation is the most efficient policy tool for generating the revenue to cover these costs. If, on the other hand, society believes each individual should be responsible for bearing his or her own burdens from natural hazards, then some form of insurance with variable rates based on the risks involved is likely to be viewed as an appropriate means of covering disaster costs.

Equity refers to concerns with fairness and the distribution of resources. An equitable distribution of resources may require the special treatment of certain individuals or groups at the expense of others. What may be viewed as equitable immediately after a disaster can be inefficient from a longer-term perspective if more people move into harm's way. For example, if uninsured disaster victims are guaranteed grants and low-interest loans that enable them to continue to locate their property in hazard-prone areas, and more people build in those areas, taxpayers will be subject to increasingly larger expenditures for bailing out victims of future disasters.

If private insurance is to play a central role in a hazard management program, as we feel it should, then the answers to the two questions posed above can be answered simply: Those in hazard-prone areas need to bear a substantial cost of making their communities safer and should .be responsible for most of the losses after a disaster occurs. The larger the subsidy provided by the general taxpayer, the less important the role that private insurance can play in covering damage from disasters. In the long run, subsidies to special groups can only be successfully achieved and maintained by some form of social insurance. If there is genuine public concern with the increasing costs of natural disasters, as we believe there is today, then an insurance system with rates based on risk can serve as the cornerstone of a hazard management program.

This book assesses the role that private and public insurance has played in the past in dealing with natural disasters and suggests how it can play a more constructive role in the future. Our position is that the economic costs of natural disasters to the nation are too high and are likely to soar in the future unless some steps are taken to change recent trends. Insurers can address these problems in a constructive manner *only* through joint efforts with other stakeholders, and through the use of strategies that combine insurance with monetary incentives, fines, tax credits, well-enforced building codes, and land use regulations. For example, one way to reduce future losses is to utilize insurance with well-enforced building codes and land use regulations.

The next section describes why natural disasters are of increasing concern to society, the nature of the available insurance, and the institutional arrangements that connect insurance to other concerned stakeholders. The section of this chapter entitled "Policy Implications" describes current public policy on natural disasters, the choices to be made by the public and private sectors in dealing with natural hazards, and the role that insurance can play in both reducing potential losses and providing compensation to disaster victims. The concluding section of this chapter describes the contents of this book.

AN OVERVIEW OF THE NATURAL DISASTER PROBLEM

An individual living in a hazard-prone area views a natural disaster as a low-probability, high-consequence event. Major hurricanes and earthquakes occur in specific regions, and then only infrequently, but when they do, the damage can be devastating, as evidenced by recent events such as Hurricane Andrew in 1992 and the Northridge earthquake in 1994. From the vantage point of a much larger geographical area such as a state or the nation, the probability of some type of natural disaster increases significantly. Federal agencies that deal with disasters, such as the Federal Emergency Management Agency (FEMA), are almost certain to incur sizable expenditures in a given year, but cannot predict the total cost or how the payout will be allocated among various types of disasters.

Increased Insured Disaster Losses

Since 1989 insurance and reinsurance firms have suffered losses from disasters in the United States that have wreaked havoc with their balance sheets. Figure 1-1 depicts the magnitude of the catastrophic losses experienced by the insurance industry in the United States from 1949 to 1997 in 1997 dollars. The drastic change from 1989 to 1997 is obvious. Prior to Hurricane Hugo in 1989 (where insured losses were over $4 billion), the insurance industry had never suffered any loss of over $1 billion from a single disaster. Since that time 10 disasters have exceeded this amount in 1997 dollars (Gary Kearney, Property Claims Services, personal communication, 1998).

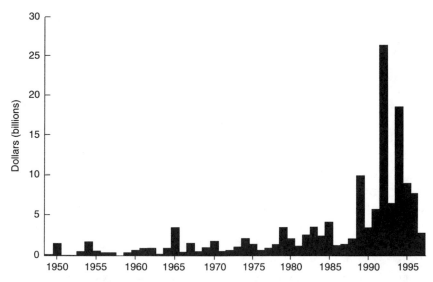

FIGURE 1-1 Insured catastrophe losses in the United States: 1949–1997 (in 1997 dollars). Source: Insurance Services Office, Inc.

Figure 1-1 sends a clear message to all concerned stakeholders that insured disaster losses have skyrocketed in recent years. The insurance industry, which was caught off guard by the very large increase in exposure in hazard-prone areas and hence had significantly underestimated the losses that could occur, is now concerned with significant natural hazards in certain parts of the United States. In these regions, especially Florida and California, insurers have prudently sought to substantially decrease their exposure and increase their rates. Some companies contend that they are facing an excessive risk of insolvency due to their high concentrations of catastrophe exposures and insufficient reinsurance (losses from Hurricane Andrew triggered the failure of nine small and medium insurers). Some also have expressed a concern that a fire sale of insurer assets (bond, stock portfolios, and real estate holdings) to pay claims from a large catastrophe could have severe repercussions for financial markets and further hamper the ability to recover. Others such as Cochrane, 1997, think that the indirect economic losses from a large-scale natural disaster would be relatively minor.

On a more positive note, new scientific studies, engineering analyses, and advances in information technology (IT) offer an opportunity to estimate the risks and potential losses of future disasters more accurately than in the past. More sophisticated risk assessments have reduced the uncertainty associated with estimating the probabilities that earthquakes and hurricanes of different intensities and magnitudes will occur in specific regions. Engineering studies, building on the experience of past disasters, have provided new information on how structures perform under the stress of natural forces. The development of faster and more powerful computers enables one to combine these data in ways that were impossible even five years ago. These new developments should give a more complete picture of property at risk so that insurers can more accurately specify premiums that reflect their expected future losses.

As for the residents of hazard-prone areas, property owners today are experiencing greater difficulty in acquiring insurance and are paying considerably more for it than they have in the past. Moreover, these market conditions will become much worse if insurers continue to withdraw from the marketplace. Consumers have found it difficult to adapt to the rapid and severe changes in insurance market conditions following recent disasters. Overbuilding and high real estate prices in the Atlantic and Gulf Coast hurricane-prone areas have, until recently, been facilitated by relatively inexpensive and readily available property insurance. As shown in Chapter 3, the majority of home buyers also appear to have ignored catastrophe risks by failing to purchase earthquake and flood coverage. Commercial development has followed the population's movement to coastal areas, and this has increased the potential economic losses from natural disasters.

The resulting requests for significantly higher insurance rates and the desire to reduce writings threaten to deflate property values and hurt economies in high-risk areas. More specifically, many residents living in very hazard-prone areas may not be able to afford insurance premiums based on the actual risk. If forced to purchase coverage in order to maintain a mortgage, the residents may have to sell their houses. Purchasers will offer a lower price than in the past because of the higher insurance costs they will be forced to pay. Residents who are uninsured will request disaster assistance following the next catastrophic event. Chapter 9 proposes a hazard management program that addresses this issue.

This situation, which we term the *natural disaster syndrome*, poses an additional challenge to those concerned with the problem of reducing future losses. Lack of interest in protection against hazards prior to a large loss places a significant financial burden on society, property owners, the insurance industry, and municipal, state, and federal governments when these losses occur. More specifically, most homeowners, private businesses, and the public sector do not adopt cost-effective mitigation measures to reduce potential losses from future disasters. A significant amount of damage would be averted if wind and seismic building codes were adopted and enforced, and if individuals took protective measures in advance of possible disasters. Serious consideration should also be given to reducing damage through more effective land use control measures.

This lack of interest in and enforcement of protective measures, coupled with the substantial growth of population in disaster-prone areas, has increased the probability that the losses will be severe when an earthquake, hurricane, or flood occurs. Given the large increase in the magnitude of losses from recent disasters, insurers and reinsurers are concerned about their financial ability to cover claims from future natural catastrophes.

Multiple Stakeholders

The key interested parties concerned with natural disaster damage, their principal roles, and their linkages with each other are depicted in Figure 1-2. At the top of the figure are the reinsurance industry, capital markets, and federal government agencies, each of which has a special role to play with respect to providing protection against catastrophic losses. The Federal Emergency Management Agency (FEMA) and other federal agencies have stressed the importance of building codes and enforcement of regulations to reduce losses from natural disasters. Reinsurers relate to insurers in the same manner that insurers relate to property owners. They provide protection to primary insurers by insuring a portion of their claims in exchange for a premium. For all but the largest insurance companies, reinsurance is a prerequisite to offering insurance against natural disasters when there is a potential for catastrophic losses.

Recently the capital markets have provided private insurers access to funds in the form of catastrophic bonds. The insurer borrows from investors or an institution at higher than normal interest rates to cover extreme losses from hurricanes and earthquakes that exceed a trigger amount. If this amount is exceeded, then the interest on the bond, the principal, or both, are forgiven. For more details on these catastrophic bonds and other financing and hedging instruments, see Doherty (1997).

The primary insurance companies, as shown in Figure 1-2, provide direct insurance coverage to residential and commercial sectors for losses such as those caused by fires (including fires resulting from earthquakes) and those caused by wind damage from tornadoes and hurricanes. Pri-

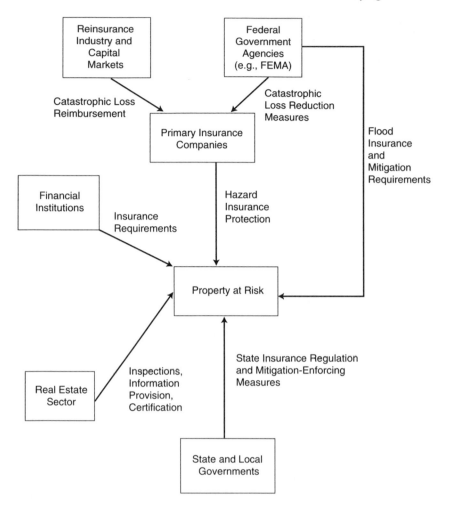

FIGURE 1-2   Role of interested parties concerned with natural disasters.

mary insurance companies offer this coverage through the standard homeowners' policies normally required as a condition for a mortgage, and through commercial multiperil policies.

Today coverage for shaking damage from earthquakes can be purchased as an addition to homeowners' policies in all states except California. A new public-private partnership, the California Earthquake Authority (CEA), formed in 1996, offers homeowners in the state earthquake coverage as a separate policy. The structure of the CEA and its impact on insurance protection of homeowners are discussed in Chapter

4. As an alternative to a CEA policy, a few private insurance companies still offer earthquake insurance as a separate policy. The type of coverage and the premiums for this insurance vary among companies. For commercial structures, earthquake protection for property damage coverage is often included as part of a multiperil policy.

Since flood hazards are considered uninsurable by private insurers, FEMA provides direct insurance protection against flood losses through the National Flood Insurance Program (NFIP). The NFIP requires flood insurance and imposes hazard mitigation requirements on all properties with federally insured mortgages in flood-prone areas. The history of the NFIP and its current status are discussed in Chapter 6.

States have the principal regulatory authority over the primary insurance companies by virtue of the McCarran-Ferguson Act (Public Law 15, 15 *U.S. Code,* sec. 1011 et seq.) passed in 1945. Each state has an insurance official who has two primary areas of responsibility: (1) monitoring and overseeing the financial solvency of the insurers, and (2) examining insurers' rates and market practices. The National Association of Insurance Commissioners (NAIC), through its advisory recommendations, plays a key role in state regulators' efforts to coordinate and strengthen their oversight of the insurance industry (Klein, 1995).

"Property at Risk," in Figure 1-2, consists of businesses, homes, and rental units located in hazard-prone areas. Property owners and tenants can reduce their potential losses from future natural disasters by adopting mitigation measures, and can protect themselves financially against the consequences of future damage by purchasing insurance and renewing their policies over time. Several other interested parties play key roles in the design and enforcement of insurance and damage mitigation requirements: financial institutions, through specific requirements as a condition for a mortgage; the construction industry, by designing safer structures; state and local governments, through ordinances and building codes; and the real estate sector, through provision of information on hazards to potential buyers and owners.

POLICY IMPLICATIONS

A coherent strategy has not yet been developed for coping with the changing role that natural hazards are playing in our lives. There is general agreement that it is important to take steps to reduce losses from future disasters. In fact, FEMA introduced a National Mitigation Strategy in December 1995 with the objective of strengthening partnerships between all levels of government and the private sector to ensure safer communities. This strategy was developed with input from state and local officials as well as individuals and organizations with expertise in hazard mitigation (FEMA, 1997).

A parallel effort has been undertaken by the Central U.S. Earthquake Consortium through a pilot project in two communities which emphasized the importance of local and business sector involvement in mitigation activities (Central U.S. Earthquake Consortium, 1997). Success in these endeavors requires the identification of cost-effective mitigation measures as well as increased expenditures to make structures safer in hazard-prone areas. To date, property owners have shown little inclination to incur such expenditures voluntarily, so there is a need for incentives, including insurance premium reductions, to encourage the adoption of these measures. In addition, well-enforced building codes are also needed to reduce collateral losses sustained in addition to those suffered by the property owner, such as damage to other structures and social costs arising from property damage (Kunreuther, 1997).

Current Disaster Policy

Current natural disaster policy places a large financial burden on all taxpayers after a disaster occurs. Under the Stafford Disaster Relief and Emergency Assistance Act of 1988 (P.L. 100-707, 42 *U.S. Code* sec. 5121 et seq.), the federal government provides funds to cover at least 75 percent of the costs of rehabilitating public facilities (U.S. Congress, 1995). For catastrophic events such as Hurricane Andrew in 1992 and the Mississippi floods of 1993, the federal government covered the entire cost of public facility repairs; it paid 90 percent of these costs after the Northridge earthquake, with the remainder financed by the State of California. Thus, it is not surprising that there has been little interest by most municipalities in investing in loss reduction measures for their facilities; city officials probably assume damage will be covered by federal or state funding. For the same reason, a number of local governments have shown little interest in adopting mitigation measures or purchasing insurance against losses to their buildings (Burby, 1992). A study by French and Rudholm (1990) of the damage to public property in the Whittier Narrows (California) earthquake of October 1987 revealed that few public buildings were protected by earthquake insurance, even though it was readily available from the private sector.

With respect to the private sector, the federal government offers assistance to uninsured and underinsured disaster victims through a Small Business Administration (SBA) disaster loan program. Under the current arrangement, homeowners and businesses suffering damage from a disaster can obtain a low-interest loan to aid their recovery. The interest rate varies between 4 percent and 8 percent. These programs can be costly to taxpayers if many such loans are provided at below-market rates. During the period from 1977 through 1993, the SBA loaned approximately $21 billion to disaster victims (U.S. Congress, 1995).

Public and Private Sector Roles

One question that needs to be addressed concerns the appropriate roles of the private and public sectors in financing the cost of recovery from large-scale natural disasters. To the extent that private insurance markets provide protection against catastrophe risk, policymakers must decide how these markets should be regulated. They must also determine the role of land use regulations and building codes and the extent to which private choice and incentives will guide hazard mitigation efforts. Within the realm of public choice, decisions also must be made with respect to the delegation of authority among the different levels of government and its agencies. In evaluating these options policymakers must consider how various government actions affect the behavior of firms and individuals in responding to catastrophe risk.

The choice between public mechanisms and private markets for financing catastrophe risk depends on how specific measures affect the decision processes of the various interested parties as well as the final outcomes. For example, it is very difficult to enforce significant cross-subsidies between different groups of risks (if that is what policymakers want to achieve) through the regulation of private insurance markets. Private businesses operating in a competitive market have strong economic incentives to avoid selling their products and services at below cost. Over time they can usually find ways to avoid or minimize their volume of such business, or to withdraw from those markets entirely. In order to implement a policy whereby rates of low-risk policyholders are used to subsidize those of higher risk, it would very likely be necessary to have legislation providing that banks and financial institutions require individuals and firms to purchase catastrophe insurance as a condition for a mortgage, and for some government agency to provide this insurance.

There is also a need to coordinate government policies affecting disaster risk. Specifically, in analyzing government policies toward insurance we must consider the impact of other private and public sector activities on hazard mitigation and the funding of recovery following a disaster. For example, the reliance on post-flood assistance and the failure of financial institutions to enforce flood insurance requirements for federally guaranteed home mortgages have adversely affected the National Flood Insurance Program's ability to encourage the purchase of flood insurance. Amendments to the NFIP in 1994 penalize banks if they do not enforce these insurance requirements. These amendments provide for a maximum of $350 for each violation, subject to a maximum calendar year amount of $100,000 for any single regulated lending institution or enterprise. As a result, there has been a substantial increase in the number of NFIP flood insurance policies purchased.

Market Failures and Government Action

The principal problems related to natural disasters are the risk of large and uncertain economic losses, the lack of availability of adequate or desired insurance coverage for some risks, the high cost of coverage, and insurance company financial difficulties and potential insolvency. In a perfectly competitive market the price of coverage should reflect the degree of risk, with a normal markup for administrative costs. However, if some of the conditions for competitive markets are not present—such as adequate information, barrier-free entry and exit, and unrestricted flow of capital—the result could be failures in the private market.

Incomplete information and misperceptions about the risk of natural hazards could cause property owners to demand a suboptimal amount of insurance coverage. Uncertainty about the true risk of loss could cause insurers to refuse to supply coverage or charge very high prices for catastrophe insurance. If insurers fail to correctly distinguish between risk types, for whatever reason, informed individuals in high-risk areas will be more inclined to purchase coverage, while those in less hazard-prone areas will be disinclined to buy coverage, resulting in a skewing of the market and an unprofitable book of business for insurers.

Entry barriers can restrict the supply of insurance, decrease competition, and raise prices. Exit barriers can further discourage entry of new insurers. Restrictions on the flow of capital also can limit the supply of coverage in places where there is considerable demand for insurance protection. Government expenditures, taxes, and regulations can create perverse incentives which distort market decisions and cause inefficiency. The challenge is to determine how government intervention can diminish or offset these market failures and increase economic efficiency and equity. The nature and effects of insurance market failures and government intervention are discussed further in the chapters to come.

ABOUT THIS BOOK

Figure 1-3 provides a road map and guide to what follows in this book.

Chapter 2 focuses on the supply side of the market by examining the insurability conditions for natural disasters. Private insurers are reluctant to continue providing widespread coverage against earthquakes and hurricanes as they have in the past. New institutions such as the California Earthquake Authority and the Florida Joint Underwriting Authority were formed in these two states because of the severe losses from the Northridge earthquake and Hurricane Andrew, respectively. Chapter 2 also examines the roles of the reinsurance industry, state insurance pools, and the capital markets in offering protection against large disasters that threaten to reduce insurers' surpluses to unacceptably low levels.

Chapter 3 focuses on the behavior of those in need of protection—residents in hazard-prone regions—and on the changing demand for residential disaster insurance in the United States. Prior to 1990 residents in hazard-prone areas had limited interest in voluntarily purchasing such coverage. Following the spate of recent disasters, the demand for insurance protection increased, as evidenced by the number of residents purchasing earthquake insurance in certain parts of California following the Loma Prieta earthquake of 1989 and Northridge in 1994. One open question concerns how price sensitive consumers are in purchasing coverage. More specifically, what impact would a rise in earthquake, wind, or flood insurance prices in certain parts of the country have on voluntary demand for coverage?

The next three chapters examine the role of insurance and other policy instruments in relation to three different types of disasters. Chapter 4 provides a history of earthquake insurance, with a focus on California, the state most concerned about and affected by that hazard. Chapter 5 examines the challenges insurers face in providing coverage against wind damage from hurricanes, using Florida as a case study. Chapter 6 reviews the changes in the National Flood Insurance Program since its establishment as a federal program in 1968 and describes its current status.

Chapter 7 examines the role that insurance and mitigation can jointly play in reducing future losses. It gives special emphasis to building codes

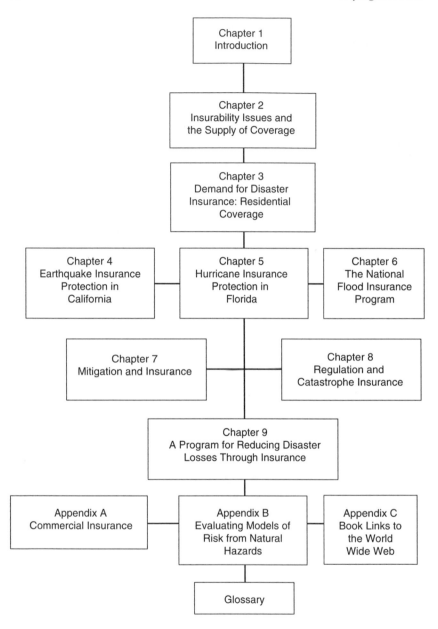

FIGURE 1-3   Road map for this book.

and their enforcement through the assistance of financial institutions and the building industry. Chapter 8 is devoted to the appropriate roles of the state insurance regulator with respect to setting rates, imposing coverage requirements, and minimizing the probability that firms will become insolvent following a catastrophic disaster.

Chapter 9 proposes a hazard management program for reducing disaster losses and providing money for recovery. The success of such an effort depends on the ability of insurers to work closely with other interested parties such as financial institutions, public sector agencies, and the capital market. This concluding chapter explores the potential role of new financial instruments, as well as some type of federal reinsurance to cover a portion of insurer and reinsurer losses from a very large catastrophe. At the end of each section of the final chapter we raise a set of open issues and questions for future research.

The three appendixes supplement the book with information on the availability of commercial insurance for natural disasters (Appendix A), the state of the art of modeling the risks from natural hazards (Appendix B), and links to information on the World Wide Web (Appendix C). These Internet linkages will enable the reader to stay up-to-date on current activities by interfacing with the many sites that are listed. A glossary at the end of the book defines technical and insurance-related terms for the general reader.

CHAPTER TWO

Insurability Conditions and the Supply of Coverage

HOWARD KUNREUTHER[1]

A S POINTED OUT IN CHAPTER 1, insurance has the potential to help reduce future losses as well as provide compensation following a disaster. However, in order for companies to be willing to supply coverage against specific risks, they must view them as being insurable. This chapter focuses on the insurance world and what it means for a risk to be insurable. It explores the challenges the insurer faces when dealing with catastrophic losses, such as those from natural disasters.

We first give a brief history of hazard insurance, paying special attention to losses from fire and natural hazards. This will make it clearer why catastrophe risks present a set of special problems for both insurers and reinsurers. We then focus on the process for determining when a risk is insurable and what premium to charge for coverage against ambiguous risks with catastrophic potential. The next section summarizes the types of insurance available for protecting residential and commercial property against types of

---

[1]James Ament, Robert Klein, Eugene Lecomte, Frank Nutter, Richard Roth, Jr., and Richard Roth, Sr., provided contributions to this chapter.

natural disasters. The chapter concludes with a discussion of how catastrophic losses have affected the insurance and reinsurance industry. In
particular, we discuss residual market mechanisms, which have been created by some states to offer insurance in high-hazard areas when the
voluntary market refuses to meet the demand at the regulated rates.

A BRIEF HISTORY OF HAZARD INSURANCE

Long before the development of insurance as such, early civilizations
evolved various kinds of insurance-like devices for transferring or sharing risk. For example, early trading merchants are believed to have
adopted the practice of distributing their goods among various boats,
camels, and caravans, so each merchant would sustain only a partial loss
of his goods if some of the boats were sunk or some of the camels and
caravans were plundered. This concept survives today in the common
insurance practice of avoiding over-concentration of properties in any
one area, and by spreading risk through reinsurance arrangements.

We can trace to the Code of Hammurabi in 1950 B.C., the practice of
bottomry, whereby the owner of a ship binds the ship as security for the
repayment of money advanced or lent for the journey. If the ship is lost
at sea the lender loses the money, but if the ship arrives safely, he receives
with the loan repayment the premium specified in advance, which is
higher than the legal rate of interest. Bottomry, one of the oldest forms
of insurance, was used throughout the ancient world (Covello and
Mumpower, 1985).

By 750 B.C. the practice of bottomry was highly developed, particularly in ancient Greece, with risk premiums reflecting the danger of the
venture. Mutual insurance also developed at this time, whereby all parties to the contract shared in any loss suffered by one of the traders who
paid a premium. The decay and disintegration of the Roman Empire in
the fifth century A.D. was followed by the development of small, isolated,
self-sufficient, and self-contained communities. Since international commerce practically ceased, there was little need for sophisticated risk-sharing devices such as insurance. However, with the revival of international
commerce in the thirteenth and fourteenth centuries, the use of insur-
ance-type mechanisms resumed (Pfeffer, 1966). In England, the establishment of Lloyd's Coffee House in 1688 (the beginnings of "Lloyd's of
London") provided a gathering place for individual marine underwriters
(Covello and Mumpower, 1985). London then became the center of the
global marine insurance market.

*Fire mark, ca. 1800, of the Philadelphia Contributionship for the Insurance
of Houses from Loss by Fire, the oldest fire insurance company in America.
The company was founded by Benjamin Franklin and his colleagues in 1752.
This plaque on a building indicated that the property had fire insurance
coverage provided by The Contributionship (Courtesy of The Philadelphia
Contributionship for the Insurance of Houses from Loss by Fire; Will Brown).*

Fire and Windstorm Insurance

Modern fire insurance was developed after the Great Fire of London
in 1666, which destroyed over three-fourths of the buildings in the city.
In the early eighteenth century Philadelphia took the first steps toward
the protection of property from the fire peril; the city possessed seven
fire-extinguishing companies. Philadelphia also passed regulations concerning the nature and location of buildings in the city. Led by Benjamin
Franklin, Philadelphians organized the first fire insurance company in

America, the Philadelphia Contributionship for the Insurance of Houses from Loss by Fire. It was followed by the creation of four more insurance companies before the end of the eighteenth century, and by an increasing number of fire insurance companies in other cities in the first half of the nineteenth century. The financial problems associated with catastrophic losses are well illustrated by the great New York fire of 1835, which swept most of the New York fire insurance companies out of existence (Oviatt, 1905).

Until recent years, attention to the fire peril has predominated over attention to other natural hazards. Insurers over the decades have studied the peril of fire intensively. Conflagrations such as the great New York fire of 1835 and the Chicago fire of 1871 focused attention on fire prevention. The National Board of Underwriters, established in 1866 in New York City, concentrated on protecting the interests of fire insurance companies. It established safety standards in building construction and worked to repress incendiarism and arson. In 1894, fire insurance companies created the Underwriters Laboratories as a not-for-profit organization to test materials for public safety. The National Fire Protection Association was founded two years later in 1896 and, by encouraging proper construction, still plays a major role in lessening the number of fires. Today's engineering and technical schools conduct research and educate and train fire safety personnel.

The nineteenth century saw the rise of the two principal types of insurance organizations in business today: mutual and stock insurance companies. A mutual company is an insurance company that is owned and operated by its policyholders and run for their benefit. The policyholders elect a board of directors, who in turn elect officers to manage the company. Because a mutual company is owned by the policyholders, any excess income is either returned to the policyholders as dividends, used to reduce premiums, or retained to finance future growth. A stock insurance company is a corporation owned by stockholders who participate in the profits and losses of the company.

The concept of insurance as an important means of risk spreading and risk reduction is best exemplified by the factory mutual insurance companies founded in the early nineteenth century in New England (Bainbridge, 1952). Factory mutual companies offered factories an opportunity to pay a small premium in exchange for protection against potentially large losses from fire, thus illustrating the risk-spreading function of insurance. One of the first mutual insurance companies providing fire insurance coverage to textile mills was Boston Manufacturers.

Following a mill inspection in 1865, Edward Manton, president of Boston Manufacturers, noted "Renew at same if an additional force pump is added. If not, renew for $10,000 at $1^1/4$" (Bainbridge, 1952, p. 112). That is, the mill would have to pay an additional $1^1/4$ cents per $100 without the additional force pump.

Regarding risk reduction, the mutuals required inspections of factories both prior to issuing a policy and after one was in force. Poor risks had their policies canceled; premiums were reduced for factories that instituted loss prevention measures. The Boston Manufacturers also worked with lantern manufacturers to encourage them to develop safer designs and then advised their policyholders that they had to purchase lanterns from those companies whose products met their specifications. The Manufacturers Mutual in Providence, Rhode Island, developed specifications for fire hoses and advised mills to buy only from companies that met these standards. In many cases, factory mutual insurance companies would only provide coverage to companies that adopted specific loss prevention methods. For example one company, the Spinners Mutual, only insured risks where automatic sprinkler systems were installed.

Insurance against wind damage was first written in the second half of the nineteenth century, but only by a limited number of companies. In its earliest form, the policy contract provided for insurance against loss by fire or storm. Until 1880 this type of insurance was sold by local or farmers' mutual insurance companies on rural risks, rather than city risks, and was confined to the East. The first insurance covering windstorm by itself was initiated in the Midwest in the latter part of the nineteenth century and was called "tornado insurance." It continued as a separate policy until 1930.

Windstorm insurance began to evolve in its present form in 1930 when the stock fire insurance companies filed an "Additional Hazards Supplemental Contract" in conjunction with the standard fire policy. This new contract provided a single rate for coverages that had traditionally been offered as separate policies for tornado, explosion, riot and civil commotion, and aircraft damage insurance. Incorporating this array of coverage in one contract was seen as a radical step in a generally conservative industry. In October of that same year, the contract was expanded even further when the New England Insurance Exchange revised it to include the perils of hail and motor vehicles, as well as all types of windstorms, not just tornadoes. In 1938, the Additional Hazards Supplemental Contract was re-named "Extended Coverage," and smoke damage was added to the coverage.

When first introduced in the East, the Extended Coverage (EC) endorsement was purchased by few individuals, and was even viewed as a luxury. A striking demonstration of this was the fact that comparatively little of this insurance was sold in Massachusetts until after the 1938 hurricane, the first to hit New England in a century. This event stimulated sale of the EC endorsement in part because many banks now required that it be added to fire insurance on mortgaged property. Although there were some changes in the perils covered during the 1930s and 1940s, the basic package has been retained and continues to be used today.

Property-Casualty Insurance Today

The insurance industry today is divided into two broad areas: property-casualty insurance companies, and life and health insurance companies. Property-casualty companies predominate numerically, with almost 4,000 property-casualty companies and about 2,000 life insurers operating in 1997 (telephone conversation, Customer Service Department, A. M. Best and Co., 1998). Property-casualty and life and health are really two different worlds, and the insurance laws of the various states recognize the sharp distinction between them. The property-casualty companies are the main source of coverage against natural hazards in the United States and hence bear the lion's share of the insured losses from disasters.

There is a significant difference between the property lines and the casualty lines, both of which are written by property-casualty companies. Property insurance reimburses policyholders directly for their insured losses, and thus is labeled as first-party coverage. Casualty insurance, for the most part, protects its policyholders against financial losses involving third parties, and consequently is called third-party insurance. For example, an automobile liability policy pays for damage caused by negligent action on the part of its insured when a vehicle owned by a third party is involved. An exception to this is automobile physical damage insurance, which is first-party insurance. It is more efficient for insurers to write this in conjunction with automobile liability coverages. General liability and workers' compensation are other important casualty lines of insurance.

Most of the financial losses caused by natural disasters are due to property damage. A homeowner's multiperil policy provides coverage against damage to one's property from fire, windstorm, and hail, as well as other perils, except for earthquake (which can be attached as a special rider to the homeowner's policy), and flood (which can be purchased from the federal government's National Flood Insurance Program by homeowners facing this risk). A commercial multiperil policy, which covers business risks, is similar to a homeowner's policy except that it often includes earthquake and flood coverage.

Agents or brokers sell most insurance policies. An agent is someone who legally represents the insurer and has the authority to act on the company's behalf. In property insurance there are two types of agents: independent agents, who can represent several companies, and exclusive agents, who represent only a single company. Brokers legally represent the insured, normally a corporation. They are extremely important in commercial property insurance today, providing risk management and loss control services to companies as well as arranging their insurance contracts. Insurance is also marketed by direct writers, who are employees of the insurer, and by mail-order companies. With the emergence of electronic commerce, the World Wide Web is likely to be an increasingly important source for marketing insurance in the future.

BASIC CONCEPTS OF INSURANCE

Insurance is an economic institution that allows the transfer of financial risk from an individual to a pooled group of risks by means of a two-party contract. The insured party obtains a specified amount of coverage against an uncertain event (e.g., an earthquake or flood) for a smaller but certain payment (the premium). Insurers may offer fixed, specified coverage or replacement coverage, which will take into account the increased cost of putting the structure back into its original condition. Law and ordinance insurance covers the additional cost required by changes in building codes and other legal requirements.

Most insurance policies have some form of deductible, which means that the insured party must cover the first portion of their loss. For example, a 10 percent deductible on a $100,000 earthquake policy means that the insurance company is only responsible for property damage that exceeds $10,000 up to some prespecified maximum amount, the coverage limit.

Losses and Claims

The insurance business, like any other business, has its own vocabulary. A *policyholder* is a person who has purchased insurance. The term *loss* is used to denote the payment that the insurer makes to the policyholder for the damage covered under the policy. It is also used to mean the aggregate of all payments in one event. Thus, we can say that there was a "loss" under the policy, meaning that the policyholder received a payment from the insurer. We may also say that the industry "lost" $12.5 billion dollars in the Northridge earthquake.

A *claim* means that the policyholder is seeking to recover payments from the insurer for damage under the policy. A claim will not result in a loss if the amount of damage is below the deductible, or subject to a policy exclusion, but there will still be expenses in investigating the claim. Even though there is a distinction between a claim and a loss, the terms are often used interchangeably to mean that an insured event occurred, or in reference to the prospect of having to pay out money.

The Law of Large Numbers

Insurance markets can exist because of *the law of large numbers,* which states that for a series of independent and identically distributed random variables (such as automobile insurance claims), the variance of the average amount of a claim payment decreases as the number of claims increases. Consider the following gambling example. If you go to Las Vegas and place a bet on roulette, you are expected to lose a little more than 5 cents every time you bet $1. But each time you bet you will either win or lose whole dollars. If you bet ten times, your average return is your net winnings and losses divided by ten. According to the law of large numbers, the average return will converge to a loss of 5 cents per bet. The larger the number of bets, the closer the average loss per bet will be to 5 cents.

Fire is an example of a risk that satisfies the law of large numbers since its losses are normally independent of one another. (In some cases, however, losses caused by fire are not independent of each other: the Oakland fire of 1991 destroyed 1,941 single-unit dwellings and damaged 2,069 others.) To illustrate this law's application, suppose that an insurer wants to determine the accuracy of the probability of fire loss for a group of identical homes valued at $100,000, each of which has a 1/1,000 annual chance of being completely destroyed by fire. If only one fire occurs in each home, the expected annual loss for each home would be $100 (i.e., 1/1000 × $100,000). If the insurer issued only a single policy, then a variance of approximately $100 would be associated with its expected annual loss. (The variance for a single loss L with probability p is $Lp (1 - p)$. If $L = \$100,000$ and $p = 1/1,000$, then $Lp (1 - p) =$ $\$100,000 (1/1,000)(999/1,000)$, or $99.90.)

A single-family house fire (FEMA).

As the number of policies issued, n, increases, the variance of the expected annual loss or mean will decrease in proportion to n. Thus, if $n = 10$, the variance of the mean will be approximately $10. When $n = 100$ the variance decreases to $1, and with $n = 1,000$ the variance is $0.10. It should thus be clear that it is not necessary to issue a large number of policies to reduce the variability of expected annual losses to a very small number if the risks are independent.

However, natural hazards—such as earthquakes, floods, hurricanes, and conflagrations such as the 1991 Oakland fire—create problems for insurers because the risks affected by these events are not independent. They are thus classified as *catastrophe risks*. If a severe earthquake occurs in Los Angeles, there is a high probability that many structures will be damaged or destroyed at the same time. Therefore, the variance associated with an individual loss is actually the variance of all of the losses that occur from the specific disaster. Due to this high variance it takes an extraordinarily long history of past disasters to estimate the average loss with any degree of predictability. This is why seismologists and risk assessors would like to have databases of earthquakes, hurricanes, or other similar disasters over 100- to 500-year periods. With the relatively short period of recorded history, the average loss cannot be estimated with any reasonable degree of accuracy.

One way that insurers reduce the magnitude of their catastrophic losses is by employing high deductibles, where the policyholder pays a fixed amount of the loss (e.g., the first $1,000) or a percentage of the total coverage (e.g., the first 10 percent of a $100,000 policy). The use of coinsurance, whereby the insurer pays a fraction of any loss that occurs, produces an effect similar to a deductible. Another way of limiting potential losses is for the insurer to place caps on the maximum amount of coverage on any given piece of property.

An additional option is for the insurer to buy reinsurance. For example, a company might purchase a reinsurance contract that will cover any aggregate insured losses from a single disaster that exceeds $50 million up to a maximum of $100 million. Such an excess-of-loss contract could be translated as follows: the insurer would pay for the first $50 million of losses, the reinsurer the next $50 million, and the insurer the remaining amount if total insured losses exceeded $100 million. An alternative contract would be for the insurer and reinsurer to share the loss above $50 million, prorated according to some predetermined percentage.

Probable Maximum Loss and Capacity

Because their financial resources are limited, insurance companies need to quantify their potential loss from a catastrophic event. In other words, insurers need an estimate of what the actual total damage might be from their current portfolio of policies and how it would affect their ability to pay claims.

For many years, fire insurance managers used the concept of a *probable maximum loss* (PML) to estimate what percentage of a particular building would likely be damaged in the event of a fire. The California Insurance Department carried this concept over to earthquake insurance when it devised its *Earthquake Questionnaire* in the 1970s (California Insurance Department, 1975). In the questionnaire the replacement cost of the insured homes is multiplied by a "PML percentage factor" to give a dollar estimate of the expected average damage to all of the insured homes in a defined earthquake zone. A similar concept is now used for defining PML in hurricane-prone areas of the country. An insurer's *capacity* is the maximum amount of PML exposure on all building risks that an insurer is willing to insure in any one region or zone. For example, when a limit is placed on a certain earthquake zone, such as $250 million, this is called a capacity limit of $250 million for that zone. Sometimes it is expressed in terms of the state as a whole. In other words, capacity is the maximum amount of aggregate loss that the insurer is willing to accept from one specific disaster event. The insurer determines its capacity based on the amount of surplus it has, its cash flow and profits from other lines of insurance it writes, and the amount of its reinsurance to cover catastrophic losses from specific disasters. Thus "capacity" refers to how much of the insurer's resources its management is willing to risk on one major disaster.

INSURABILITY CONDITIONS[2]

What does it mean to say that a particular risk is insurable? This question must be addressed from the vantage point of the potential supplier of insurance who offers coverage against a specific risk at a stated premium. The policyholder is protected against a prespecified set of losses defined in the contract.

Two conditions must be met before insurance providers are willing to offer coverage against an uncertain event. Condition 1 is the ability, when providing different levels of coverage, to identify and quantify, or estimate, the chances of the event occurring and the extent of losses likely to be incurred. Condition 2 is the ability to set premiums for each potential customer or class of customers. This requires some knowledge of the customer's risk in relation to others in the population of potential policyholders. If Conditions 1 and 2 are both satisfied, a risk is considered to be insurable. But it still may not be profitable. In other words, it may be impossible to specify a rate for which there is sufficient demand and incoming revenue to cover the development, marketing, and claims costs of the insurance and still yield a net positive profit. In such cases the insurer will opt *not* to offer coverage against this risk.

Condition 1: Identifying the Risk

To satisfy this condition, estimates must be made of the frequency at which specific events occur and the magnitude of loss they are likely to cause. Such estimates can use data from previous events, or scientific analyses of what is likely to occur in the future. We will first illustrate the use of historical data, with the peril of fire as an example, and then discuss scientific prediction, with the earthquake peril as an example.

---

[2]This section is based on Chapter 4 in Freeman and Kunreuther, 1997.

We will then show how new developments in catastrophe risk modeling have made it easier to estimate the risk facing insurers when they are considering how much coverage to provide in hazard-prone areas.

Collecting Historical Data on the Risk of Fire

Rating agencies typically collect data on all the losses incurred over a period of time for particular risks and exposure units. Suppose the hazard is fire and the exposure unit is a well-defined entity, such as a $300,000 wood frame home to be insured for one year in California. The typical measurement is the pure premium (PP), which is the basis for setting an actual premium for potential customers. The PP is defined as:

$$PP = \text{Total Losses/Total Number of Exposure Units. (1)}$$

(The pure premium normally considers loss adjustment expenses for settling a claim. We will assume that this component is part of total losses. For more details on calculating pure premiums, see Launie et al., 1986.) Assume that the rating agency has collected data on 100,000 wood frame homes in that state and has determined that the total annual loss from fires to these structures over the past year is $20 million. If these data are representative of the expected loss to this class of wood frame homes in California next year, then, using (1), PP is given by:

$$PP = \$20,000,000/100,000 = \$200.$$

This figure is simply an average. It does not differentiate between wood frame homes of different values, their locations in the state, the distance of each home from a fire hydrant, or the quality of the fire department serving different communities. All of these factors are often taken into consideration by underwriters and actuaries who specify a premium that reflects their estimate of the risk to particular structures.

Using Scientific Data to Estimate the Risk of Earthquakes

If there were comparable long-term data available on annual damage to wood frame homes in California from earthquakes of different magnitudes, then a method similar to the one described above could be used to determine the probability and magnitude of loss from earthquakes. However, due to the infrequency of these disasters and the relatively small number of homes that have been insured against them, this type of analysis is not feasible. Insurance providers instead have turned to scientific studies by seismologists, geologists, and structural engineers to estimate the frequency of earthquakes of different magnitudes, as well as the damage that is likely to occur to different structures from such earthquakes.

Figure 2-1 depicts the type of information required to determine the pure premium for a wood frame house subject to earthquake damage in California. The *x*-axis (magnitude of loss) is the amount of damage an earthquake might cause to a wood frame home of a given value. The *y*-axis (probability) is the estimated annual probability that a wood frame home in a specific region of California would suffer a specified amount of loss from an earthquake. If these data are available from scientific studies, the pure premium in this case would be equivalent to the expected loss, which is given by the area under the curve in Figure 2-1.

Scientists have been working to reduce the ambiguity and uncertainty in predicting the location, severity, frequency of occurrence, and physical effects of earthquakes. Over the past 20 years seismologists have discovered factors that influence the probability of an earthquake

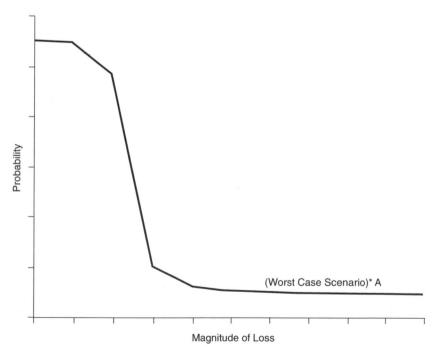

FIGURE 2-1 Determining expected loss to a wood frame house from an earthquake in California.

in a specific area, by examining geologic records, looking at actual events, and conducting experiments on how the ground responds to earthquake processes. Scientists are still uncertain as to how different factors interact with each other and what their relative importance is (Hanks and Cornell, 1994).

Engineers have focused on the nature, distribution, and level of damage from earthquakes. Congress mandated the continued improvement of worldwide post-earthquake investigations by authorizing the National Earthquake Hazard Reduction Program in 1976, and re-authorizing it in 1990 (P.L. 101-614). Such investigations have increased our understanding of the performance of various types of buildings and structures in earthquakes of different magnitudes. However, there is still considerable uncertainty about the damage earthquakes are likely to cause to different structures. Hazard risk maps have been drawn for earthquakes, but they only provide rough guidelines as to the likelihood and potential damage from specific events. For the medium-intensity Northridge earthquake, for example, the predicted damage was considerably less than the actual losses (Scawthorn, 1995). (For a detailed discussion of new advances in seismology and earthquake engineering, see FEMA, 1994, and Office of Technology Assessment, 1995.)

The recent use of geographic information systems (GIS) for incorporating geologic and structural information for a region has enabled scientists to estimate potential damage and losses from different earthquake scenarios. The data for the region are stored in the form of GIS maps of ground shaking estimation; maps of secondary seismic hazards such as liquefaction, landslide, and fault rupture; and maps of damage to structures in the region (King and Kiremidjian, in press).

While seismologists and geologists cannot predict with any certainty the probability of earthquakes of different magnitudes occurring in specific regions of California and the resulting damage to structures, they are often asked to predict worst-case scenarios for a particular structure as a result of an earthquake of a given magnitude. For example, they are asked to estimate the maximum credible probability (p^*i) from an earthquake of intensity (i) in the vicinity of a wood frame house in an earthquake-prone area. The engineers then provide their best estimates of the probable maximum loss to the house (L^*_i) from such an earthquake. (The use of PML for determining earthquake premiums is described in Chapter 4.) The resulting worst-case scenario for the wood frame house is shown as point A in Figure 2-1.

Development of Catastrophe Models

New advances in information technology have led to the development of catastrophe models, which have proven very useful for quantifying risks based on estimated probabilities and expected damage. The development of faster and more powerful computers now makes it possible to examine extremely complex phenomena in ways that were impossible even five years ago. Large databases can now be stored and manipulated so that large-scale simulations of different disaster scenarios under various policy alternatives can be easily undertaken.

A catastrophe model is the set of databases and computer programs designed to analyze the impact of different scenarios on hazard-prone areas. A catastrophe model combines scientific risk assessments of the hazard with historical records to estimate the probabilities of disasters of different magnitudes and the resulting damage to affected structures and infrastructure (U.S. Congress, 1995). The information can be presented in the form of expected annual losses and/or the probability that in a given year the claims will exceed a certain amount. Catastrophe models can also be used to calculate estimated insured losses from specific hypothesized events (e.g., a Hurricane Andrew hitting downtown Miami and Miami Beach).

Specifically, catastrophe models combine the characteristics of the disaster with characteristics of the property in the affected region to determine a damageability matrix. This matrix provides information on the potential losses from disasters of different magnitudes to the structures at risk. Depending on the type of insurance coverage available, one can then estimate the insured loss per property (see Insurance Services Office, 1996). Figure 2-2 illustrates the interaction between these different components of the catastrophe model for earthquake hazards. A more detailed discussion on how one creates and utilizes catastrophe models appears in Appendix B.

The occurrence of Hurricane Andrew and the Northridge earthquake stimulated the insurance industry to pay more attention to output from these catastrophe models that indicate what could happen in hurricane- and earthquake-prone areas over periods of 10 years, 100 years, or even 1,000 or 10,000 years. Catastrophe models provide the insurer with an opportunity to determine how much coverage it should provide, what premiums it should charge, and where it should offer policies to reduce its probability of severe financial losses to an acceptable level.

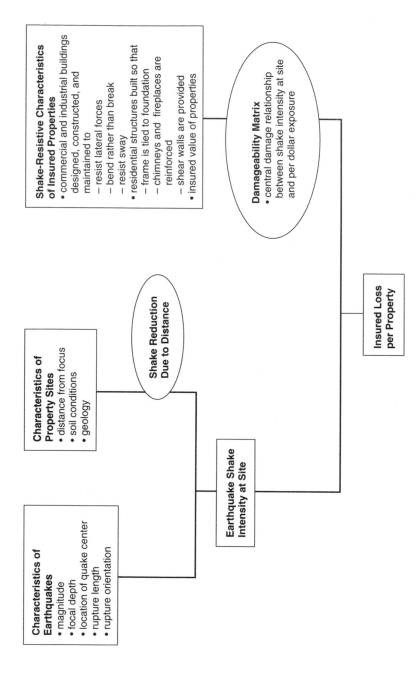

FIGURE 2-2 Modeling the effects of earthquake shake damage on insured losses. Source: Insurance Services Office, 1996.

Condition 2: Setting Premiums for Specific Risks

Once the risk has been identified, the insurer needs to determine what premium it can charge to make a profit, while not subjecting itself to an unacceptably high chance of a catastrophic loss. There are a number of factors that play a role in determining what prices companies would like to charge. In the discussion that follows we are assuming that insurers are free to set premiums at any level they wish. In reality, state regulations often limit insurers in their rate-setting process.

Ambiguity of Risk

Not surprisingly, the higher the uncertainty regarding the probability of a specific loss and its magnitude, the higher the premium will be. As shown by a series of empirical studies, actuaries and underwriters are so averse to ambiguity and risk that they tend to charge much higher premiums than if the risk were well specified.

Kunreuther et al. (1995) conducted a survey of 896 underwriters in 190 randomly chosen insurance companies to determine what premiums would be required to insure a factory against property damage from a severe earthquake. The survey results examine changes in pricing strategy as a function of the degree of uncertainty in probability and/or loss. A probability is considered to be well specified where there is enough historical information on an event that all experts can agree that the probability of a loss is p. When there is wide disagreement about the estimate of p among the experts, this ambiguous probability is referred to as Ap. L represents a known loss—that is, there is a general consensus about what the loss will be if a specific event occurs. When a loss is uncertain, and the experts' estimates range between L_{min} and L_{max}, this uncertain loss is denoted as UL. Combining the degree of probability and loss uncertainty leads to four cases which are shown in the columns of Table 2-1.

To see how underwriters reacted to different situations, four scenarios were constructed, as shown by the rows in Table 2-1. Where the risk is well specified, the probability of the earthquake is either .01 or .005; the loss, should the event occur, is either $1 million or $10 million. The premium set by the underwriter is standardized at 1 for the non-ambiguous case; one can then examine how ambiguity affects pricing decisions.

Table 2-1 shows the ratio of the other three cases relative to the

TABLE 2-1   Ratios of Underwriters' Premiums for Ambiguous or Uncertain Earthquake Risks Relative to Well-Specified Risks[a]

Scenario	Cases				
	1 p,L	2 Ap,L	3 p,UL	4 Ap,UL	5 N^b
$p = .005$ $L = \$1$ million	1	1.28	1.19	1.77	17
$p = .005$ $L = \$10$ million	1	1.31	1.29	1.59	8
$p = .01$ $L = \$1$ million	1	1.19	1.21	1.50	23
$p = .01$ $L = \$10$ million	1	1.38	1.15	1.43	6

[a]Ratios are based on mean premiums across number of respondents for each scenario.
[b]N = number of respondents.

Source: Kunreuther et al., 1995.

nonambiguous case (p,L) for the four different scenarios, which were distributed randomly to underwriters in primary insurance companies. For the highly ambiguous case (Ap,UL), the premiums were between 1.43 to 1.77 times higher than if underwriters priced a nonambiguous risk. The ratios for the other two cases were always above 1, but less than the (Ap,UL) case.

Adverse Selection

If the insurer sets a premium based on the average probability of a loss, using the entire population as a basis for this estimate, those at the highest risk for a certain hazard will be the most likely to purchase coverage for that hazard. In an extreme case, the poor risks will be the only purchasers of coverage, and the insurer will lose money on each policy sold. This situation, referred to as *adverse selection*, occurs when the insurer cannot distinguish between the probability of a loss for good- and poor-risk categories.

The assumption underlying adverse selection is that purchasers of insurance have an informational advantage by knowing their *risk type*. Insurers, on the other hand, must invest considerable expense to collect information to distinguish between risks. For example, suppose some homes have a low probability of suffering damage (the good risks) and others have a higher probability (the poor risks). The good risks stand a 1 in 10 probability of loss and the poor risks, a 3 in 10 probability. For simplicity, assume that the loss is $100 for both groups and that there are an equal number of potentially insurable individuals in each risk class.

Since there is an equal number in both risk classes, the expected loss for a random individual in the population is $20. (This expected loss is calculated as follows: [50(.1 × $100) + 50(.3 × $100)]/100 = $20.) If the insurer charges the same premium across the entire population, only the poor-risk class would normally purchase coverage, since their expected loss is $30 (.3 × $100), and they would be pleased to pay only $20 for the insurance. The good risks have an expected loss of $10 (.1 × $100), so they would have to be extremely risk averse to be interested in paying $20 for coverage. If only the poor risks purchase coverage, the insurer will suffer an expected loss of $10 (i.e., $30 − $20) on every policy it sells.

There are two principal ways that insurers can deal with this problem. If the company knows the probabilities associated with good and bad risks but does not know the characteristics of the property, the insurer can raise the premium to at least $30 so that it will not lose money on any individual purchasing coverage. In reality, where there is a spectrum of risks, the insurer may only be able to offer coverage to the worst-risk class in order to make a profit. Hence, raising premiums is likely to produce a market failure in that very few of the individuals who are interested in purchasing coverage to cover their risk will actually do so at the going rate.

A second way for the insurer to deal with adverse selection was proposed by Rothschild and Stiglitz (1976). The insurer could offer two different price-coverage contracts. For example, Contract 1 could be offered at price = $30 and coverage = $100, while Contract 2 might be price = $10 and coverage = $40. If the poor risks preferred Contract 1 over 2, and the good risks preferred Contract 2 over 1, this would be one way for the insurers to market coverage to both groups while still breaking even.

A third approach is for the insurer to require some type of audit or examination to determine the nature of the risk more precisely. However, inspections and audits are expensive and will raise the premium charged unless the potential policyholder pays for the audit.

It is important to remember that the problem of adverse selection only emerges if the persons considering the purchase of insurance have more accurate information on the probability of a loss than the firms selling coverage. If the policyholders have no better data than the insurers, both sides are on an equal footing. Coverage will be offered at a single premium based on the average risk, and both good and poor risks will want to purchase policies.

Moral Hazard

Providing insurance protection to an individual may lead that person to behave more carelessly than before he or she had coverage. If the insurer cannot predict this behavior and relies on past loss data from uninsured individuals to estimate rates, the resulting premium is likely to be too low to cover losses.

Moral hazard refers to an increase in the probability of loss caused by the behavior of the policyholder. Obviously, it is extremely difficult to monitor and control behavior once a person is insured. How do you monitor carelessness? Is it possible to determine if a person will decide to collect more on a policy than he or she deserves by making false claims?

The numerical example used above to illustrate adverse selection can also demonstrate moral hazard. With adverse selection the insurer cannot distinguish between good and bad risks, but the probability of a loss for each group is assumed not to change after a policy is sold. With moral hazard the actual probability of a loss becomes higher after a person becomes insured. For example, suppose the probability of a loss increases from $p = .1$ before insurance to $p = .3$ after coverage has been purchased. If the insurance company does not know that moral hazard exists, it will sell policies at a price of $10 to reflect the estimated actuarial loss ($.1 \times \$100$). The actual loss will be $30 since p increases to .3. Therefore, the firm will lose $20 (i.e., $30 − $10) on each policy it sells.

One way to avoid the problem of moral hazard is to raise the premium to $30 to reflect the known increase in the probability, p, that occurs once a policy has been purchased. In this case there will *not* be a decrease in coverage as there was in the adverse selection example. Those individuals willing to buy coverage at a price of $10 will still want to buy a policy at $30 because they know that their probability of a loss with insurance will be .3.

Another way to avoid moral hazard is to introduce *deductibles* and *coinsurance* as part of the insurance contract. A sufficiently large deductible can act as an incentive for the insureds to continue to behave carefully after purchasing coverage because they will be forced to cover a significant portion of their loss themselves. With coinsurance the insurer and the insured share the loss together. An 80 percent coinsurance clause in an insurance policy means that the insurer pays 80 percent of the loss (above a deductible), and the insured pays the other 20 percent. As with a deductible, this type of risk-sharing arrangement encourages safer behavior because those insured want to avoid having to pay for some of the losses. (For more details on deductibles and coinsurance in relation to moral hazard, see Pauly, 1968.)

Another way of encouraging safer behavior is to place upper limits on the amount of coverage an individual or enterprise can purchase. If the insurer will only provide $500,000 worth of coverage on a structure and contents worth $1 million, then the insured knows he or she will have to incur any residual costs of losses above $500,000. This assumes that the insured will not be able to purchase a second insurance policy for $500,000 to supplement the first one and hence be fully protected against a loss of $1 million, except for deductibles and insurance clauses. Even with these clauses in an insurance contract, the insureds may still behave more carelessly with coverage than without it simply because they are protected against a large portion of the loss. For example, insureds may decide not to take precautionary measures that they would have adopted had they been uninsured. The cost of adopting mitigation may now be viewed as too high relative to the dollar benefits that the insured would receive from this investment. If the insurer knows in advance that an individual will be less interested in loss reduction activity after purchasing a policy, then it can charge a higher insurance premium to reflect this increased risk, or it can require specific mitigation measures as a condition of insurance. In either case this aspect of the moral hazard problem will have been overcome.

Correlated Risk

Correlated risk refers to the simultaneous occurrence of many losses from a single event. As pointed out earlier, natural disasters such as earthquakes, floods, and hurricanes produce highly correlated losses: many homes in the affected area are damaged and destroyed by a single event.

If a risk-averse insurer faces highly correlated losses from one event, it may want to set a high enough premium not only to cover its expected losses, but also to protect itself against the possibility of experiencing catastrophic losses. An insurer will face this problem if it has many eggs in one basket—if, for example, the insurer provides earthquake coverage mainly to homes in Los Angeles County rather than across the entire state of California.

To illustrate the impact of correlated risks on the distribution of losses, assume that there are two policies sold against a risk, where $p =$ .1, and $L = \$100$. The actuarial loss for each policy is $10. If the losses are perfectly correlated, then there will be either two losses with probability of .1, or no losses with a probability of .9. On the other hand, if the losses are independent of each other, then the chance of two losses decreases to .01 (i.e., $.1 \times .1$), with the probability of no losses being .81 (i.e., $.9 \times .9$). There is also a .18 chance that there will be only 1 loss (i.e., $.9 \times .1 + .1 \times .9$).

The expected loss for both the correlated and uncorrelated risks is $20. [For the correlated risk the expected loss is $.9 \times \$0 + .1 \times \$200 =$ $20. For the independent risk the expected loss is $(.81 \times \$0) + (.18 \times \$100) + (.01 \times \$200) = \$20$.] However, the variance will always be higher for correlated than uncorrelated risks if each has the same expected loss. Thus, risk-averse insurers will always want to charge a higher premium for the correlated risk.

Insurability Conditions and Demand for Coverage

The above discussion suggests that in theory insurers can offer protection against any risk that they can identify, and for which they can obtain information to estimate the frequency and magnitude of potential losses, as long as they have the freedom to set premiums at any level. However, due to problems of ambiguity, adverse selection, moral hazard, and highly correlated losses, they may want to charge premiums that considerably exceed the expected loss. For some risks the desired premium may be so high that there would be very little demand for coverage at that rate. In such cases, even though an insurer determines that a particular risk meets the two insurability conditions discussed above, it will not invest the time and money to develop the product.

More specifically, the insurer must be convinced that there is sufficient demand to cover the development and marketing costs of the coverage through future premiums received. If there are regulatory restrictions that limit the price insurers can charge for certain types of coverage, then companies will not want to provide protection against these risks. In addition, if an insurer's portfolio leaves the insurer vulnerable to the possibility of extremely large losses from a given disaster due to adverse selection, moral hazard, and/or high correlation of risks, then the insurer will want to reduce the number of policies in force for these hazards.

The case studies of California and Florida discussed in Chapters 4 and 5 reveal a number of concerns by insurers with respect to providing coverage against earthquake and hurricane. Given rate and coverage restrictions imposed on them in each of these states, insurers claim that insurance for these two perils *cannot* be profitably marketed today. (It should be noted that most insurance companies made large profits prior to the early 1990s by marketing hurricane insurance in Florida and earthquake coverage in California. However, all of these profits, and more, disappeared with the occurrence of Hurricane Andrew and the Northridge earthquake.) Insurers had reached a similar conclusion a number of years ago with respect to flood hazards. This led to the development of the National Flood Insurance Program in 1968.

AVAILABLE RESIDENTIAL COVERAGE
FOR NATURAL DISASTERS

We now turn to the types of coverage that are currently available in the United States for protection against property damage from different types of natural disasters.

Earthquakes

As pointed out in Chapter 1, insurance coverage for earthquake shake damage can be included, but is not automatically included, in homeowners' insurance policies for an additional premium, except in California. Today in California, one normally purchases a residential earthquake policy through the California Earthquake Authority, a state-run earthquake insurance company. A more detailed analysis of earthquake insurance in California appears in Chapter 4. For states including California, earthquake coverage is often included in a commercial policy for structures in hazard-prone areas, or it can also be purchased as a separate policy. Protection against loss by fire which might follow an earthquake is included in the basic fire peril coverage in all property insurance contracts. Business interruption (BI) from an earthquake is covered by a separate BI policy.

Floods

Flood insurance was first offered by private companies in the late 1890s and again in the mid-1920s (Manes, 1938). However, the loss experienced by those insurers providing coverage was so large that they left the market and discouraged others from entering. Insurance companies viewed the flood risk as unmarketable because problems of adverse selection and high correlation of risks necessitated rates so high that few individuals would want to purchase coverage. The 1952 *Report on Floods and Flood Damage* explicitly makes this point by noting that because of the "impossibility of making this line of insurance self-supporting due to the refusal of the public to purchase such insurance at the rates which would have to be charged to pay annual losses, companies generally could not prudently engage in this field of underwriting" (Insurance Executive Association, 1952).

In 1968 Congress created the National Flood Insurance Program (NFIP) as a means of offering coverage on a nationwide basis through the cooperation of the federal government and the private insurance industry. A more detailed discussion of the NFIP, and its opportunities and challenges in comparison with a private insurance program, is discussed in Chapter 6.

Hailstorms

Hail is a product of thunderstorms, and is usually associated with strong winds. Since damage is caused by both wind and hail, it is difficult to allocate the dollar loss to each cause separately. Thus, insurance loss payment data is combined for the two perils. Insurance coverage for hailstorms is nearly universal. It is part of the basic coverage in virtually all residential and commercial property insurance policies sold in the United States. Anyone purchasing a property insurance policy in this country probably has the coverage.

Although hailstorms can produce widespread structural damage, it is unusual for them to render a dwelling uninhabitable or a business structure unusable. Hail is the second leading cause of damage covered by property insurance in the United States, after fire, and the variance in the aggregate level of losses from year to year is relatively small in contrast to other natural hazards.

Hailstorm damage can occur virtually anywhere in the country, although some areas are considerably more susceptible to these storms. Because hailstorms are widespread and quite random, it is possible to forecast expected loss costs with a reasonable degree of certainty. This, in turn, allows property insurers and reinsurers to price this coverage and allocate capital with sufficient accuracy to fund expected loss costs.

Hurricanes

Hurricanes are large storms that originate over tropical ocean waters; they are characterized by high winds and extensive flooding. Hurricane wind damage and windblown water damage are included as part of the basic wind coverage in most property insurance policies. Flood damage resulting from hurricanes is *not* included in property insurance policies but can be purchased as separate coverage under the National Flood Insurance Program.

Although relatively infrequent in the United States, when hurricanes do occur they are capable of causing tremendous destruction and damage to structures over a large area. Property exposures to hurricanes in all coastal states (with the exception of Louisiana) more than doubled from 1980–1993. During this period the value of insured residential exposures increased by 166 percent, and commercial exposures went up 193 percent. (Insurance Research Council and Insurance Institute for Property Loss Reduction, 1995). Since these storms are events of low frequency but high financial impact, considerable uncertainty surrounds loss projections for the future. Chapter 5 presents a detailed discussion of the challenges that insurers face in providing hurricane coverage in Florida.

In recent years increased attention has been given to hurricane prediction. Historical data for the twentieth century suggest that hurricanes follow cyclical patterns. After a relatively quiescent period from 1960 to 1987, the picture changed when Hurricane Gilbert hit the Caribbean and Mexico in 1988, and Hurricane Hugo struck the South Carolina coast in 1989. In 1992 Hurricane Andrew devastated south Florida, and Hurricane Iniki hit Hawaii. In 1995, Hurricane Erin and Hurricane Opal caused significant damage in Florida.

Landslides

Landslides are normally not considered an insurable peril by private insurers. Insurance programs cover landslides only to the extent that the damage is caused by an insured earthquake, or is associated with flood damage covered by policies sold under the National Flood Insurance Program.

Lightning

While lightning is not usually considered a disaster peril, it is a natural cause of loss that results in a number of damage events in the United States. Lightning damage is covered by the basic fire peril and, as such, is included in standard property insurance policies.

Most lightning damage is relatively minor, consisting of damage to electrical components caused by a lightning surge traveling through a structure's electrical system. Occasionally, fire will result from a lightning strike. Lightning storms seldom produce a large number of serious damage claims in a single event. Like tornadoes and hailstorms, these events are sufficiently random that they seldom result in market disruptions.

Tornadoes

Like hailstorms, insurance coverage for tornado damage is nearly universal. Coverage for tornado damage is part of the basic wind coverage in all residential and commercial property insurance policies. The pricing and marketing characteristics of tornado coverage are similar to those of hailstorms.

Tornadoes are possible throughout the United States, but historical data indicate that some states are more tornado-prone than others. Tornado alley stretches from west Texas to Kansas, Oklahoma, and the rest of the midwestern United States. Most of the 900 tornadoes recorded annually in the United States occur in this vicinity. Texas has the greatest number, with almost 100 touching down each year (Harris et al., 1992). Since the national weather service began tracking tornadoes in 1953, the three states with the most tornado sightings have been Texas, Oklahoma, and Florida, in descending order (Grazulis, 1993).

Tornadoes often cause severe damage or destruction to structures directly in their paths, and less serious damage to structures on their periphery. Typical tornadoes involve relatively small areas of damage, although there are notable exceptions such as the ones hitting Lubbock, Texas (1970), and Xenia, Ohio (1974), each of which caused damage in the hundreds of millions of dollars. Since tornadoes result from the same kinds of meteorological conditions that produce hailstorms, they are often accompanied by hail damage.

Tsunamis

The public often calls tsunamis (or seismic sea waves) "tidal waves." This implied relationship to ocean tides is incorrect, as tsunamis are directly related to earthquakes; tides are the result of gravitational forces of the moon. One of the most disastrous tsunamis of the last century occurred on June 15, 1896, in the Sankriku region on the northeast coast of Japan. About 20 minutes after feeling a slow gentle motion of long duration (the indicators of a large distant shock), the sea receded to return as a wall of water tens of feet high, killing 27,122 people.

Wildfires

Wildfires can best be defined as fires fed by natural vegetation that cause damage to structures and other property in their path. Included in that definition are forest fires, prairie fires, and brush fires (more frequently referred to as urban wildfires in recent years).

Wildfires are sufficiently random to permit the private insurance mechanism to include this coverage in most property insurance contracts. There are, however, instances where the probability of loss is so great that insurance coverage is not readily available through private carriers. Examples include remote areas that are susceptible to wildfires but have minimal firefighting services, and areas with particularly hazardous physical surroundings such as the brush areas of southern California. In some of these regions, state governments have mandated insurance pools to insure these properties. Any losses incurred by these pools in excess of premiums paid are assessed to private carriers marketing insurance in the state.

Winter Storms

Severe winter storms produce structural damage and destruction. Typical damages are caused by high winds, weight of ice and snow, freezing damage to plumbing systems, and fires resulting from heating units operating beyond their designed capacity. In many instances, this damage results from storms in areas that do not normally experience severe winter weather. While seldom resulting in total damage to individual structures, these events can produce widespread moderate damage with large aggregate costs.

Damage caused by winter storms is included in most property insurance policies because losses from these events are sufficiently random to permit the private insurance mechanism to operate effectively.

Volcanic Eruption

Damage directly caused by volcanic eruption is included in most property insurance policies. Not much property in the United States is exposed to this type of hazard.

ALTERNATIVE MECHANISMS FOR FINANCING CATASTROPHIC LOSSES

The worst catastrophe-related insured losses in the United States, by a wide margin, occurred during the period from 1989 to 1997, as shown in Table 2-2. The realization that potential damage is as much as five to six times greater than had been previously assumed greatly impacts the amount of capital considered at risk by an insurer. In the extreme case, an insurer that had intended to limit its catastrophe exposure to the amount of its retained capital (net worth) would now find that its exposure is several times greater than anticipated. This insurer would have no choice but to reduce that exposure.

The more likely circumstance is a multiline, multistate insurer that sets its disaster risk "appetite" at a relatively small fraction of retained capital. When that insurer learns that its real exposure significantly exceeds that amount, it must also adjust. Basically, that adjustment can be accomplished in three ways: the amount of exposure can be reduced, the amount of capital can be increased, or some type of protection can be provided to the insurer against catastrophic losses from specific disasters. This section focuses on the third method of providing financial relief, showing how private reinsurance and state residual market arrangements have offered insurers added protection against catastrophic losses.

The Role of Reinsurance

Reinsurance does for the insurance company what primary insurance does for the policyholder or property owner: it provides a way to protect against unforeseen or extraordinary losses. For all but the largest insurance companies, reinsurance is almost a prerequisite for offering insurance against hazards where there is the potential for catastrophic damage. In a reinsurance contract, one insurance company (the reinsurer, or assuming insurer) charges a premium to indemnify another insurance company (the ceding insurer) against all or part of the loss it may sustain under its policy or policies of insurance.

TABLE 2-2 Annual Catastrophe Losses, 1949–1997

Year	Number of Catastrophes	Estimated Loss Payments
1949	4	$ 22,300,000
1950	12	231,150,000
1951	5	17,450,000
1952	9	24,050,000
1953	16	88,650,000
1954	10	298,600,000
1955	12	94,900,000
1956	11	72,400,000
1957	11	73,350,000
1958	7	25,300,000
1959	7	48,200,000
1960	8	129,400,000
1961	15	183,500,000
1962	20	197,000,000
1963	9	33,700,000
1964	21	196,150,000
1965	13	694,100,000
1966	16	110,800,000
1967	37	327,222,000
1968	22	134,509,000
1969	20	256,408,000
1970	21	450,162,000
1971	35	173,458,000
1972	30	214,704,000
1973	41	376,076,000
1974	31	696,040,000
1975	32	513,492,000
1976	27	270,745,000
1977	44	422,653,000
1978	41	645,563,000
1979	54	1,702,554,000
1980	51	1,177,023,000
1981	33	713,831,000
1982	33	1,528,670,000
1983	33	2,254,765,000
1984	26	1,548,258,000
1985	34	2,816,035,000
1986	26	871,516,000
1987	24	946,000,000
1988	32	1,409,000,000
1989	34	7,642,000,000
1990	32	2,825,000,000
1991	36	4,723,000,000
1992	36	22,970,000,000

continued

TABLE 2-2   *Continued*

Year	Number of Catastrophes	Estimated Loss Payments
1993	36	$ 5,620,000,000
1994	38	17,010,000,000
1995	34	8,325,000,000
1996	41	7,375,000,000
1997	25	2,600,000,000
Total 1949–1988	933	$21,989,684,000
Total 1989–1997	312	$79,090,000,000

Note: Figures not adjusted for inflation.

Source: Property Claims Service.

The most common type of reinsurance contract is a *treaty,* a broad agreement covering some portion of a particular class of business. (A less common form of reinsurance is a facultative contract, which covers a specific risk of a ceding insurer and is often written for a business where there is a significant potential for loss, such as airlines.) There are several types of treaties, each of which involves different sharing arrangements between the insurer and reinsurer. To illustrate these differences, consider the hypothetical case where a large fire causes $10 million in damages to a policyholder of the Accord Insurance Company. Under the terms of a *quota share treaty* arranged with the Binder Reinsurance Company, the two companies would agree to share losses according to some fixed percentage. If the two companies share losses equally, Accord and Binder Reinsurance would each pay $5 million in claims.

Under a *pro rata treaty* Accord would pay the entire claim and be reimbursed for a set percentage of the claims by Binder Re in exchange for an annual premium. If Binder Re had agreed to pay 30 percent of the loss, then Accord initially pays the entire $10 million and receives $3 million back from the reinsurer.

If the arrangement between the companies had been an *excess of loss treaty*, Accord would be responsible for all losses up to a specified amount, and Binder Re would reimburse Accord for a layer of losses up to some prespecified maximum dollar figure. For example, if the reinsurance contract specified $5 million in excess of $1 million, then Accord would pay the $10 million in losses, and Binder Re would reimburse Accord for $5 million of this amount. If the insured loss was below $1 million, then Accord would be responsible for all of it.

Reinsurance is a global business. Its international nature reflects a desire to spread risk and access domestic and foreign capital markets to help cover losses. About two-thirds of all property catastrophe reinsurance placed on risks occurring in the United States is held by foreign reinsurance companies.

The property catastrophe reinsurance market has reacted quickly to major catastrophe losses. The increase in those losses over the past few years has alerted reinsurers to the magnitude of accumulated exposures. Worldwide reinsurance capacity suffered at the end of 1992 into 1993, not only from retrenchment by U.S. reinsurers following hurricanes Andrew and Iniki, but also because of shrinkage in the London market—a vital player in the global reinsurance business. In 1993, reinsurers significantly renegotiated contract terms in light of the reevaluation of catastrophe exposures. Today, the outlook has brightened and the market has rebounded. The Lloyd's market is coming back with new cash commitments from corporate entities and better management of aggregate exposures. In addition, market opportunity for new capital resulted in the establishment of the "Bermuda Market" for property catastrophe reinsurance. Eight new property catastrophe reinsurance companies were formed between June 1993 and January 1, 1994, with combined capital of $4.8 billion.

The recent trend of consolidation among reinsurers, which has developed over the past few years, bodes well for the property catastrophe reinsurance market. As of the end of 1997, catastrophe reinsurance capacity remained abundant. Extrapolating from the maximum lines offered by reinsurers from different worldwide markets, Guy Carpenter and Company has estimated that the amount of traditional capacity available for any single insurance program is realistically around $500 million. An additional $500 million can be obtained from nontraditional sources (Guy Carpenter and Company, 1997).

A major factor behind market consolidation has been a "flight to quality"—that is, a demand by ceding companies for greater reinsurance security, which has translated into higher capitalization for reinsurers in general. Further, property catastrophe reinsurers that maintain financial strength through sufficient capitalization and have a geographically diversified book of business will be well positioned to respond to market demands for catastrophe reinsurance.

Residual Market Mechanisms

Residual markets mechanisms (RMMs) are designed to allow persons who are unable to obtain insurance through normal channels to purchase insurance coverage. RMMs are established in most instances by state legislation and are run by all insurance companies doing property insurance business in the state. RMMs have played an important role in enabling insurers to continue to offer coverage against losses from natural disasters, notably hurricanes. Despite the availability of some reinsurance, there is concern among the insurance companies that their current exposure to hurricanes is much too large and that they must reduce the number of policies in hazard-prone areas. Because of strict regulation of insurers by the various state insurance departments, it is not possible to cancel policies wholesale. Hence the need for RMMs.

FAIR Plans

The Fair Access to Insurance Requirements, or FAIR, plans were born as a result of the civil disorders, riots, and conflagrations that erupted in the mid- and late 1960s. These disturbances resulted in a severe shortage of fire insurance in the urban areas affected. It became apparent that without the availability of insurance, the damaged and destroyed cities would not be rebuilt. In 1968 Congress passed the Urban Property Protection and Reinsurance Act, which enabled urban property owners to have fair access to insurance as a prerequisite for obtaining building rehabilitation loans vital to the preservation of the cities.

The states, which regulate insurers and insurance, enacted laws creating FAIR plans and mandating all licensed property-casualty insurers to participate. Insurers, in turn, were protected against catastrophic losses from riots through reinsurance made available by, and purchased from, the federal government (Lecomte and Demerjian, 1997). The federal reinsurance facility never was accessed, and ceased to exist when the 1968 act finally expired in 1983. It did, however, establish a precedent for the use of FAIR plans with respect to natural hazards perils.

Beach and Wind Plans

Beach and wind plans were developed at about the same time as FAIR plans. The numerous hurricanes which ravaged the Atlantic and Gulf Coasts during the 25-year interval between 1941 and 1965 had resulted in limited availability of windstorm insurance in these hazard-prone areas. This availability problem prompted legislatures in the affected states to enact laws requiring insurers providing homeowners' coverage to make insurance against loss by windstorm available to those who otherwise would not be able to purchase a policy. Thus, beach and windstorm plans, similar to FAIR plans, evolved in most states along the Atlantic coast (Lecomte and Demerjian, 1997).

Evaluation of RMMs

In theory, RMMs are intended to serve only as a secondary source of coverage for insurable risks that cannot obtain coverage through the voluntary market. This implies that the plans charge adequate rates (i.e., do not run a deficit) and remain relatively small. However, depending on how they are structured and regulated, RMMs can significantly exacerbate market problems. These mechanisms have the potential of mushrooming and even swallowing the voluntary market if their rates are competitive, or if other factors significantly restrict the supply of voluntary coverage. High assessments to fund residual market deficits can also hasten an insurer's exit from the voluntary market and discourage entry. Cost containment incentives can be undermined if residual markets rates are inadequate, and/or servicing carriers are not induced to control claim costs.

RMMs cannot be used effectively to replace the voluntary market or substantially subsidize high-risk areas. At best, they can provide only a short-term safety valve for temporary decreases in the supply of insurance, and a long-term source of coverage for a small number of high-risk properties. The creation of these mechanisms, however, does not provide a solution for the solvency of the insurance companies because the sponsoring companies are responsible for the losses that arise from them. The net result is that the losses are merely redistributed among the companies depending upon how the assessment procedure is fashioned for the particular residual markets organization.

SUMMARY AND CONCLUSIONS

The insurance industry today is concerned with the magnitude of losses from catastrophic events because most firms perceive that another major disaster could affect their financial stability. Many insurers in the

United States would like to reduce the amount of coverage they offer and charge premiums more in line with their risk experience.

These concerns raise a question as to what the future role of the private market will be in providing financial protection against losses from catastrophic events. Many insurers feel that the premiums they are forced to charge are too low, and the number of policies they are required to provide in hazard-prone areas places them in a precarious financial position should another catastrophic disaster occur in these areas.

Because future earthquakes and hurricanes pose a threat to the solvency of insurers, new approaches are required for dealing with catastrophe risks. It is, however, essential that both insurance and non-insurance participants recognize that by far the greater amount of insurance claim payments continues to be made for non-catastrophic losses. While totals vary considerably from year to year and from company to company, over the past 35 years only about 20 percent of insurance claim payments have gone to catastrophe-related losses. This figure includes the financial impact of both Hurricane Andrew and the Northridge earthquake. The remaining 80 percent of payments have been made for routine fire, theft, liability, and other losses which occur one at a time.

However, if a disaster does occur, it may cause such a financial drain on the insurers' surplus that it may restrict their ability to provide other types of coverage. To prevent this situation, it is necessary to explore arrangements for dealing with the risks of large losses from natural disasters. The next two chapters discuss how California and Florida have addressed these issues with respect to the earthquake and hurricane risks, respectively.

CHAPTER THREE

Demand for Disaster Insurance: Residential Coverage

RISA PALM[1]

W
HY IS THERE LIMITED DEMAND for disaster insurance? If people understood that they can protect themselves from catastrophic losses with a small investment in insurance, would they purchase coverage? Several studies on how people make decisions when faced with unlikely, but possibly disastrous events provide considerable insight into why many residents in hazard-prone areas do not purchase insurance voluntarily to protect themselves against losses from future disasters (Camerer and Kunreuther, 1989). Despite the intuitive appeal of the axioms associated with normative models of choice, the plain fact is that few individuals behave according to expected utility theory—that is, most individuals do not select the strategy that maximizes expected utility. Instead, it appears that the weights on probability and outcomes in judging different alternatives are contingent on the problem context or the framing of different options (Tversky and Kahneman, 1981). What are the factors that affect the decision to purchase disaster insurance? And, more generally, how do human beings adjust to risks in their environments?

[1]Edward Pasterick provided contributions to this chapter.

FACTORS INFLUENCING THE
PROTECTION AGAINST HAZARDS

Survey research has explored the relationship between protection against specific risks and a series of factors such as income level, age, gender, and prior experience with the hazard. In general, empirical research has found that lower-income persons, non-white individuals, women, the elderly, and persons with previous disaster experience will show greater fear of the hazard. In some, but only some, of these cases this fear will also be translated into the adoption of measures to avoid losses, such as buying disaster insurance. Frequently survey research findings have been contradictory, and, even more frequently, have not been explained by reference to an overarching theoretical perspective (Drabek, 1986).

Another line of research, more theoretically based, goes beyond individual characteristics or experience to describe the process of decision making. The simplest theoretical framework for predicting the purchase of insurance is an analysis of the ratio of expected benefits to certain costs. In theory, it is possible to assess the probability of losses of different magnitudes for a piece of property in a given year, and then calculate the relative costs of purchasing insurance and receiving claim payments. The property owner could then compare this amount to the estimated cost of no insurance premium combined with absorbing the costs of repairs oneself should the property be damaged. For this type of cost-benefit analysis to predict not only the "ideal" practice, but also the observed practice, the decision maker must have the goal of maximizing expected utility and must possess all relevant information about costs and benefits in an uncertain world. Since these assumptions about omniscience cannot be met in actuality, this conceptualization is modified into one of subjective expected utility (Edwards, 1955; Friedman and Savage, 1948; Mosteller and Nogee, 1941).

However, even with this modification, empirical studies show that besides a lack of information, other factors prevent decision makers from undertaking even simple cost-benefit analysis. For one thing, individuals often exhibit a set of biases with respect to their estimates of the probability in making judgments under uncertainty. For example, they may be prone to the gambler's fallacy—believing that if a low-probability event has occurred recently, it is unlikely to occur again soon and therefore can be treated as an event with a probability of zero (Slovic et al., 1974).

Individuals may utilize a set of probability rules through an editing procedure to simplify the choice process. For example, decision makers may equate low probability with zero probability, which enables them to ignore certain alternatives (Kahneman and Tversky, 1979; Slovic et al., 1977). They may anchor their estimate of a loss on a salient figure and then adjust their estimate around this first approximation—an approximation that may be highly inaccurate and will bias later estimates (Einhorn and Hogarth, 1985; Tversky and Kahneman, 1974). In short, the notion that individuals calculate the costs and benefits of various alternatives and decide on some set of "rational" adaptations to the environment does not fit the empirical reality of complexity in decision making. Individuals not only lack complete knowledge of alternatives, but also demonstrate patterns of consistent "errors" in risk calculation.

Kunreuther (1996) argues that individuals tend to make different trade-offs between such issues as the probability of the event or its likely outcomes depending on the context of the problem and the means used to communicate the information. "People often weight these dimensions differently than would be suggested by normative models of choice such as expected utility theory or benefit cost analysis" (Kunreuther, 1996, p. 175). For example, some people may show little interest in adopting insurance against natural hazards because they believe that it "cannot happen to me." People may not adopt protective measures because they are myopic in their view of the future and are therefore unprepared to pay the relatively high up-front cost in relation to these perceived short-run benefits.

On the other side of the coin, the decision to purchase earthquake insurance is a specific example of what Hogarth and Kunreuther label "decision making under ignorance," where both costs and benefits are unknown to the decision maker. In such conditions, these authors argue, "people determine choices by using arguments that do not quantify the economic risks and may reflect concerns that are not part of standard choice theory" (Hogarth and Kunreuther, 1995, p. 16). Instead, people justify their decisions using arguments that may seem far-fetched or may not resemble normative models of choice.

Individual decision making may also take place in cultural contexts that constrain or enable the range of available choices. The cultural context may increase or reduce awareness of risk, and condition the range of acceptable responses. Wildavsky and Dake (1990) note that personality structures are neither risk-averse nor risk-taking in all situations, and therefore tests of such factors cannot predict hazard response. Instead, "cultural biases provide predictions of risk perceptions and risk-taking preferences that are more powerful than measures of knowledge and personality" (Wildavsky and Dake, 1990, pp. 171–172).

A specific example of the impacts of cultural context on hazard response is the research related to "optimism." A number of studies suggest that, in comparing themselves to others, Americans see themselves as living longer, as being younger for their age, and as less likely to die from cardiovascular diseases or accidents (Myers, 1992). Garrison Keilor's claim that all children of Lake Wobegon are "above average" pokes fun at this tendency. Researchers find an absence of this optimistic bias among Japanese college students (Heine and Lehman, 1995; Markus and Kitayama, 1991).

This American optimism about personal well-being could also affect perceived vulnerability and the propensity to protect oneself from risk by purchasing insurance or taking steps to protect one's home from future disasters. Empirical findings from surveys of Japanese and California homeowners showed that Californians tended to be overly optimistic, to believe their own neighborhoods were safer and better prepared for earthquakes than other areas in their city or region, while Japanese believed their own areas were more at risk and less well prepared (Palm and Carroll, 1998).

FLOOD AND EARTHQUAKE INSURANCE

Many risks stemming from the environment, broadly defined, confront homeowners, from toxic waste spills to civil disorder. This chapter focuses on two major natural hazards facing households—flooding and earthquakes—and, specifically, on the decision to purchase insurance against the damage caused by these hazards. Although wind and tornadoes can cause destruction to residential property, insurance against such damage is usually included in the general homeowners' policy, as pointed out in the previous chapter.

Empirical Evidence Concerning Flood Insurance Purchase

Floods account for more losses than any other natural disaster in the United States, and for more federal disaster assistance in most years. Riverine floods threaten every part of the United States, although damage is highly dependent on local variables.

Since 1968, flood insurance has been provided by the federal government under the National Flood Insurance Program (NFIP) through the Federal Emergency Management Agency (FEMA). (See Chapter 6 for a comprehensive description of the NFIP.)

Congress sets the rates and terms of flood insurance under the NFIP. The program specifies requirements for building in the floodplain and provides subsidized insurance rates for structures in existence before these requirements were adopted. Federally regulated or insured lending institutions offering home mortgages on property in the 100-year floodplain must require flood insurance as a condition for granting an initial mortgage loan. However, in the past, lenders have permitted mortgage holders to let their flood insurance policies lapse. A recent survey estimates a 70 to 80 percent lapse rate among policyholders after the initial years of the mortgage (KRC Research and Consulting, 1995). Further, since no fine was levied on lenders who failed to comply, there was no economic incentive for them to follow the requirement (Federal Interagency Floodplain Management Task Force, 1992). This has since been corrected by passage of the National Flood Insurance Reform Act of 1994 (P.L. 103-325, 108 *Stat.*, 2255).

During the first four years of the subsidized flood insurance program fewer than 3,000 out of 21,000 flood-prone communities entered the program, and fewer than 275,000 homeowners voluntarily bought a flood insurance policy. Insurance agents had little economic incentive to sell this separately marketed insurance since their commissions were low relative to the time required to market coverage (Kunreuther et al., 1978). By 1992, a conservative estimate of coverage suggests that less than 20 percent of the homes located in the floodplain were covered by flood insurance (Kusler and Larson, 1993). The Federal Insurance Administration estimates that as of 1997 about 27 percent of households living in high-risk flood areas are insured. This estimate is based on about 10 million total households in high-risk zones, with 2.7 million policies in force. About 1 million policies have been sold outside high-risk areas.

The NFIP requires participating communities for which flood maps have been developed to have minimum building codes and standards. Although mapping is completed in most of the communities undergoing development, those unmapped communities cannot implement these building standards (Kusler and Larson, 1993). Thus, new development continues in areas potentially at risk from flooding.

In 1995, the Federal Emergency Management Agency commissioned a study to "facilitate its efforts to sell flood insurance to the public nationwide." KRC Research and Consulting conducted 27 interviews in

August 1995 with lenders, realtors, community officials, and advisory board committee members (KRC Research and Consulting, 1995). Among other things, they found that the public still has little knowledge about flood insurance coverage, just as the Kunreuther team found 20 years earlier (Kunreuther et al., 1978). Individuals purchase policies primarily to comply with regulations rather than because of perceived risk or a desire for the insurance per se.

Empirical Evidence Concerning Earthquake Insurance Subscription

Earthquakes are also a national problem. The highest magnitude (M) earthquake on the Richter scale within North America in the twentieth century occurred in Alaska in 1964 (M = 8.4), and more earthquakes of similar magnitude are predicted for that region. Within the lower 48 states, the highest magnitude earthquake occurred near New Madrid, Missouri at the beginning of the nineteenth century (M = 9.2 in 1812). A recurrence of such an earthquake in this region would wreak havoc, particularly given the relative lack of adequate local construction standards and preparedness.

Earthquake insurance, unlike fire insurance, is not generally required by mortgage lenders as a condition for a mortgage loan. A survey of lenders conducted in 1982 (Palm et al., 1983) found that a majority of home mortgage lenders in California stated that they rarely or never make adjustments in lending terms simply because property is in a landslide-prone area, a special earthquake studies zone, or an area known to be underlain by a surface fault trace, or because of evidence of seismic damage. Lenders in California as well as the state of Washington ranked earthquake hazards as the least likely of five possible causes of mortgage defaults. They report that it is far more likely that default would follow unemployment, divorce, a house fire, or a major flood rather than an earthquake.

Lenders gave several reasons why property owners ignored the earthquake risk. First, most loans are made on post-1940 houses. These houses were built according to some seismic building code, resulting in less vulnerability to major earthquake-related damage. Second, even if the property sustained major damage, lenders believe that there is little probability that the borrowers would default on the mortgage loan as long as they have positive net equity in the home. Third, lenders believe that they spread their vulnerability over such a wide range of investments, and over a sufficiently large geographic area, that even if a major earthquake occurred in one part of the state, their portfolios would not be subject to major losses. Fourth, lenders who participate in the secondary mortgage market have passed the risk on to others. Fifth, lenders may not use geologic information for lending decisions because of state and federal anti-redlining legislation preventing geographic mortgage lending discrimination. In the 1982 survey, lenders used one or more of these arguments to justify their ignoring of seismic conditions in making loan decisions in California.

Looking at homeowners' decisions to purchase earthquake insurance, we would expect that because of the national distribution of earthquake risk there would also be a national demand for earthquake insurance. However, most of the actual demand is in the state of California, and to date the most comprehensive cross-sectional studies on the demand for disaster insurance have been undertaken in California. At the time of the first systematic survey of earthquake insurance purchase (Kunreuther et al., 1978), only five percent of homeowners were insured against this risk. Over the 20-year period since that survey, several events have occurred that impact insurance adoption: (1) passage of legislation requiring the disclosure of information on earthquake risk and on the availability of earthquake insurance, and (2) the occurrence of several moderate-scale earthquakes.

Hazard Disclosure Legislation

The two most relevant legislative acts are the Alquist-Priolo Special Studies Zones Act and a state law requiring the earthquake insurance disclosure. The Alquist-Priolo Act was passed in March 1972 in response to the San Fernando earthquake of February 1971. This act was intended to prevent new large-scale development or siting of facilities, such as hospitals and schools, in areas particularly susceptible to fault rupture. In 1975, the act was amended to require disclosure that a property was in a special earthquake studies zone: "A person who is acting as an agent for a seller of real property which is located within a delineated special studies zone, or the seller if he is acting without an agent, shall disclose to any prospective purchaser the fact that the property is located within a delineated special studies zone" (California Public Resources Code, sec. 2621.9).

The amended act required purchasers of property within one-fourth mile of an active surface fault trace to be informed about the hazard from fault rupture. The Alquist-Priolo Act was amended again in 1990 to extend the nature of the hazard zones and to mandate disclosure of these new zones. The new law required the State Mining and Geology Board to develop guidelines for the preparation of maps of seismic hazard zones by January 1, 1992, and required the seller or the agent of the seller to disclose "to any prospective purchaser the fact that the property is located within a seismic hazard or delineated special studies zone, if the maps or information contained in the maps are reasonably available." The legislation mandates the mapping of seismic hazard zones, including areas susceptible to "strong ground shaking, liquefaction, landslides, and other ground failure," and specifies additional information that must be disclosed to the potential new purchaser of property.

It is unclear how this new information is being used in the purchase process. An early study of the impacts of special studies zones disclosure on home buyer behavior showed that most buyers did not understand or remember the disclosure (Palm, 1981). The question of whether the modified hazard disclosure mandated in the 1990 legislation results in increased awareness of individual risk exposure merits future study. Because of the required disclosure to prospective property purchasers in the special studies zone, it might be expected that home buyers would translate this increased awareness of the earthquake hazard into the adoption of mitigation measures as well as the purchase of insurance. Although evidence shows an increase in demand for insurance, the specific causes have not been pinpointed.

A second form of disclosure legislation was aimed more broadly at all property owners who carry homeowners' insurance, normally required as a condition for a mortgage loan and purchased by most owners without a mortgage. Insurance companies were required to mention the availability of earthquake insurance and to make clear that fire insurance policies do not cover earthquake damage. The relevant statute (California Insurance Code, sec. 2, 1081) went into effect in 1984. It requires homeowners to be informed biannually of the availability and cost of earthquake insurance and thus might motivate some homeowners to consider purchasing this coverage.

Earthquake Experience

The occurrence of several moderate earthquakes in California between 1983 and 1994 could be expected to have influenced the purchase of earthquake insurance. People do more to prepare for future earthquakes, including buying insurance against them, just after an earthquake than at any other time. As the memory of the earthquake fades, the motivation for preparedness decreases.

Between the Kunreuther survey in 1974 and the last of the Palm surveys in 1995, six moderate-scale earthquakes occurred. The first was the 6.7 magnitude earthquake at Coalinga in 1983, causing $31 million in property damage and 205 injuries. This earthquake damaged approximately 1,000 housing units, about 40 percent of the total in the city. The 1987 Whittier Narrows earthquake had a lower magnitude (5.9) and resulted in no deaths or serious injuries, but it damaged 5,000 buildings and caused losses of approximately $358 million. The Loma Prieta earthquake of 1989 was a magnitude 7.1 earthquake, which resulted in 62 deaths, 3,000 injuries, damage to 18,300 houses, and property damage of approximately $6 billion.

In 1992, California earthquakes occurred in several locations: the Cape Mendocino earthquake near Petrolia (magnitude 7.0), which damaged numerous historic dwellings, and the Landers/Big Bear earthquake (with magnitudes of 7.6 and 6.5, respectively). The Landers/Big Bear earthquake sequence was the strongest to occur in the state since 1952. It caused 1 death and 25 injuries, and resulted in property losses to private and public buildings in excess of $90 million (Earthquake Engineering Research Institute, 1992).

The most recent damaging earthquake was the magnitude 6.7 Northridge earthquake of January 1994, which resulted in 33 fatalities, most caused by structural failure (Earthquake Engineering Research Institute, 1994). Fifteen people died in a "dingbat" style, three-story wood frame apartment building in Northridge. Although the death count in the Northridge earthquake was very low, the economic consequences of this earthquake were staggering: insured losses topped $12.5 billion (California Insurance Department, 1995), with total losses exceeding $50 billion (Risk Management Solutions, 1995a). Publicity about the disaster raised the awareness of California residents concerning their vulnerability to earthquake damage.

As a result of the increase in seismic activity in California, there was a significant demand for earthquake insurance between 1983 and 1995. In the 1970s less than 10 percent of the homes were insured against earthquake damage. By 1995 over 40 percent of the homes in many areas along the coast were insured against earthquake damage. Since the threat of earthquake damage is much less in the interior of the state, fewer homes are insured for earthquake damage in inland cities such as Fresno and Sacramento.

Survey Findings on Insurance Purchase

Three surveys of a population of owner-occupiers in Contra Costa, Santa Clara, Los Angeles, and San Bernardino Counties in 1989, 1990, and 1993 showed a dramatic increase in earthquake insurance purchase from the 1973–1974 baseline, and a gradual increase in all these counties over the four-year study period (Palm, 1995).

Contra Costa County had the lowest percentage of earthquake insurance purchase, starting at 22 percent insured in 1989 and ending with 37 percent insured in 1993, as shown in Table 3-1. The highest percentage of earthquake insurance purchase was in Santa Clara County, with the percentage of insured properties jumping more than 10 points in the single year following the Loma Prieta earthquake of 1989.

There was a major increase in San Bernardino County, also related to the Landers/Big Bear earthquake sequence. Los Angeles County respondents showed a steady increase in insurance purchase, to a slight majority insured by 1993. Overall, earthquake insurance purchase has increased. However, a large number of households—a majority in San Bernardino and Contra Costa Counties—remain uninsured. An even more comprehensive census of earthquake insurance purchase shows heavy concentrations of insured households in urbanized coastal counties where vulnerability is high.

Reasons for Insurance Purchase

In the 1989–1993 surveys, those who purchased insurance were asked to assess factors that affected their purchase decision. Figure 3-1 shows that the most important motivating factors for those who purchased insurance, in order of importance were: (1) "worry that an earthquake will destroy my house or cause major damage in the future"; (2)

TABLE 3-1 Percent of Survey Respondents Purchasing Earthquake Insurance

County	1989	1990	1993
Contra Costa	22.4	29.3	36.6
Santa Clara	40.4	50.9	54.0
Los Angeles	39.6	45.8	51.6
San Bernardino	29.2	34.6	42.6

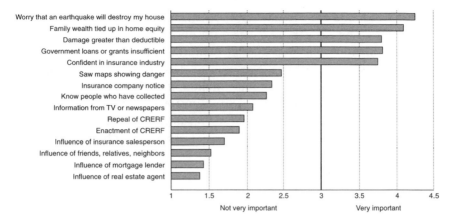

FIGURE 3-1 Factors that influenced the decision to purchase earthquake insurance, 1989–1993. Note: CRERF = California Residential Earthquake Recovery Fund (see below).

"most of our family wealth is tied up in the equity of our house, which might be lost if an earthquake destroyed or damaged it"; (3) "if a major earthquake occurs, the damage to my house will be greater than the deductible, so insurance is a good buy"; and (4) "if a major earthquake occurs the grants or loans available from the federal or state government will not be sufficient to rebuild my house." The mean score for each of these factors was at least 3.5 (where 5 was "very important" and 1 was "not at all important").

The least important reasons for purchasing insurance were "my neighbors or friends or relatives or colleagues convinced me to have earthquake insurance," "my real estate agent encouraged me to buy it," and "my mortgage lender suggested that I have it." Insurance purchase was thus motivated by anticipated losses, fear that governmental aid will be unavailable or insufficient, and an estimate of likely damages as opposed to the cost of premiums. The influence of family, friends, real estate agents, or mortgage lenders was negligible.

These findings represent a change in the factors influencing the purchase decision since the time of the Kunreuther survey of 1973–1974. In the earlier survey, knowing someone with insurance and talking about insurance with someone were among the most influential factors in causing the household to consider insurance and to purchase it. The shift from the influence of friends and neighbors to more economic motivations (the subject's own judgment about anticipated losses, potential government aid, and the probability of an earthquake) may be due to the passage from an early stage of the diffusion of an innovation to a later stage in the adoption process, demonstrating the existence of a more mature market for earthquake insurance. That is, when earthquake insurance was relatively new and uncommon, it took a personal acquaintance with an insured individual to motivate the spread of insurance among California households. At present, since so many people already have earthquake insurance, assessments of risk and cost have become the motivating factors.

Insuring the Deductible

In 1992 the State of California experimented with a fund to provide insurance of up to $15,000 to cover losses associated with the deductible on catastrophe insurance. The California Residential Earthquake Recovery Fund (CRERF) levied a surcharge of from $12 to $60 (depending on location and type of dwelling) on residential and mobile home insurance policies. Coverage of up to $15,000 would be guaranteed only for the structure, with deductibles of $1,000 to $3,500 depending on the value of the house. As pointed out in Chapter 4 (p. 73), an estimated 90 percent of California homeowners paid the surcharge for this program, but the program was repealed in September 1992.

The 1993 survey asked homeowners whether they had paid this surcharge and the reasons for their decision. Although the program was short-lived, and perhaps of only historic interest, their responses have important implications for the success of future programs, whether marketed through private insurance companies or again through governmental guarantees. In the 1993 survey, overall 59 percent of the respondents said that they had paid the surcharge. (This is less than the 90 percent participation mentioned earlier because it includes policies that went into effect after 1992 and were thus ineligible for CRERF coverage.) Contra Costa County respondents showed the lowest purchase percentage (54 percent), and Santa Clara, Los Angeles, and San Bernardino Counties ranged from 61 to 62 percent subscription to the CRERF. As depicted in Figure 3-2, most of those who paid the CRERF surcharge did so because they wanted the extra coverage and/or they thought it was mandatory.

In contrast, those who did not pay the surcharge had a variety of reasons—primarily they believed they did not need this type of policy, or they doubted the permanence or soundness of the program (see Figure 3-3).

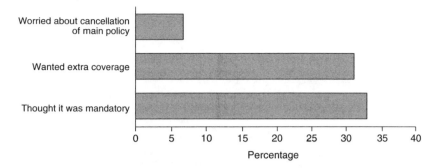

FIGURE 3-2   Why respondents paid CRERF surcharge.

These survey findings suggest that any future program, even with relatively low premiums and covering most of the potential losses, must seem dependable to the potential purchasers or the program will not be adopted.

Who Purchases Insurance?

The 1989–1993 surveys found that the propensity to buy insurance was not consistently or systematically related to the demographic or economic characteristics of the homeowners. Although factors such as income, age, and length of residence in California might be statistically associated with earthquake insurance purchase in one county or another, these variables did not, individually or collectively, distinguish the insured from the uninsured in the four counties surveyed. However, individual estimation of potential destruction of the home by an earthquake was the variable most closely associated with the purchase of insurance.

Surveys in 1994 and 1995

More recent information has been collected about insurance purchase behavior of Californians. The general purpose of this study was to compare the possible impacts of culture-based personality factors on earthquake hazard response and the adoption of mitigation measures in areas of similar hazard in California and Japan (Palm and Carroll, 1998). The California study areas—Cupertino, Redlands, and portions of the western San Fernando Valley—were selected as areas of predominantly owner-occupiers, who were also white and middle class. Table 3-2 shows

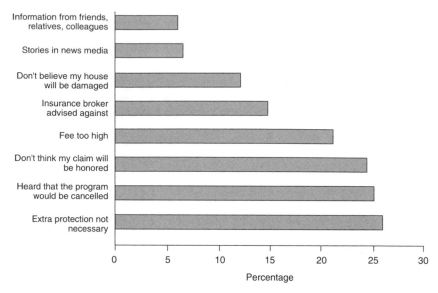

FIGURE 3-3   Why respondents did not pay CRERF surcharge.

that earthquake insurance purchase ratios were quite high in all three study areas, exceeding two-thirds of the respondents in the case of Cupertino.

As in previous studies, perceived risk was strongly related to the tendency to adopt earthquake insurance: those who were more worried about earthquakes, and who believed that their own homes were likely to be damaged by an earthquake in the next ten years, were also more likely to purchase earthquake insurance.

In order to assess the types of measures individuals and households in the study areas adopted, we presented them with a checklist derived from a composite of the various measures suggested in the *Homeowner's Guide to Earthquake Safety* developed by the California Seismic Safety Commission in response to 1991 legislation (Government Code, Title

TABLE 3-2   California Earthquake Insurance Purchase in 1995

Study Area	Percentage Who Purchased Earthquake Insurance
Cupertino	67.2
Redlands	47.7
Western San Fernando Valley	53.9

*The Northridge Meadows Apartments, where 16 people were killed in the
Northridge earthquake of 1994. The top two stories collapsed onto the first
floor, which had garages at either end, with apartments in the middle. The
apartment buildings were constructed with joist hangers that were too small,
and joists were not nailed to hangers. Shear walls were not made of plywood
but gypsum board. There was reduced shear strength on the lower floor
because of the garages. Steel supporting beams bent and collapsed
(USGS, J. Dewey).*

21, Division 1, Chapter 1.38). Other items were added or deleted in
response to suggestions by members of the advisory committee to create
a final list of eleven items ranging in cost from strengthening the house
itself to cost-free items such as making plans for reuniting the household
after an earthquake or planning an escape route.

We found that the purchase of earthquake insurance was positively
associated with certain mitigation measures taken in the home, such as
arranging heavy objects so that they were less likely to fall, participating
in earthquake drills, purchasing a fire extinguisher, and investing in measures that would strengthen the house. Earthquake insurance purchase,
however, was not related to the adoption of mitigation measures such as
preparing an escape plan, storing food and water, making plans for family reunions after the earthquake, storing items in a container for evacuation, knowing how to turn off gas and other utilities, or having a first
aid kit.

CONCLUSIONS ABOUT EARTHQUAKE
INSURANCE PURCHASE

Empirical research has demonstrated that the demand for one form of disaster insurance—insurance against catastrophic losses associated with earthquakes—has greatly increased in recent years. However, there seems to be some limit to demand for this insurance given budget constraints, current premium rates, and perceptions on the part of at least a portion of the population at risk that the probability of disaster is so small that "it cannot happen to me." Those who purchase insurance are worried about destruction from a future earthquake; those who eschew insurance believe it is too expensive, and that their houses are not susceptible to major damage. Clearly, universal, voluntary insurance coverage, even in an area at risk from earthquakes such as metropolitan southern California, is unlikely to be realized.

One should note that the survey data on earthquake coverage reported here only reflects homeowners' decisions in one area of the United States at one period of time (1989–1995). Since that time, the California Earthquake Authority has replaced the private insurance market mechanism with respect to this type of coverage in California. Chapter 4 provides more detailed evidence on the demand for insurance under this new program.

Finally, the decision processes by firms and business organizations in the private and public sectors in other parts of the United States and the world may be very different from those in California. Also, significant cultural differences may affect the decision to purchase coverage in other countries, particularly in non-Western nations.

CHAPTER FOUR

Earthquake Insurance Protection in California

RICHARD J. ROTH, JR.

C ALIFORNIA IS WELL KNOWN for having earthquakes—the 1906 San Francisco earthquake comes readily to mind—yet damage-causing earthquakes have not been that common in California until recently. After 1906, the next damage-causing earthquakes occurred in Santa Barbara (1925) and in Long Beach (1933). None of the state's building codes incorporated earthquake-resistive features until 1933, and then only for schools and public buildings. After the 1906 earthquake, builders, property owners, and government officials did not act on the dangers of earthquakes. Californians today live with their legacy: buildings of unreinforced masonry construction and homes unattached to their foundations.

THE SAN FERNANDO EARTHQUAKE

The modern era of earthquake activity began with the 1971 San Fernando earthquake, which caused substantial damage to homes, businesses, and public buildings. Insurance covered only a small portion of the losses because very few people had earthquake coverage and there was very little fire damage. (In

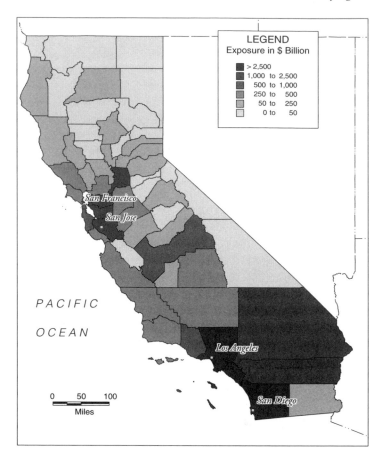

LEGEND
Exposure in $ Billion

> 2,500
1,000 to 2,500
500 to 1,000
250 to 500
50 to 250
0 to 50

San Francisco
San Jose

PACIFIC

OCEAN

Los Angeles

0   50   100

Miles

San Diego

FIGURE 4-1 Total residential and commercial exposures in California.

California as well as other states, homeowners' coverage covers any fire losses, even if caused by an earthquake.)

The 1971 disaster caused several important changes in attitudes. It raised the consciousness of people with respect to the dangers of earthquakes, and it focused attention on the necessity of revising the building codes to require earthquake-resistive design and construction. The event also reinforced the insurance industry's concern about the threat of earthquakes, and prompted the California Insurance Department to request private insurers to report annually on their earthquake exposures. This questionnaire, still in use today, is valuable to the department in monitoring the solvency of insurers. Figure 4-1 shows the distribution of current earthquake exposures in California.

THE NORTHRIDGE EARTHQUAKE

Although a series of damaging earthquakes have occurred in California since the San Fernando earthquake, none of them compare to the January 17, 1994, Northridge earthquake. Next to Hurricane Andrew, it is the largest insured natural disaster ever, causing the largest insured damage by far of any earthquake in the United States, with total insured losses of over $12.5 billion. (In contrast, only a fraction of the damage caused by the Kobe, Japan, earthquake in 1995 was covered by insurance.) To put the $12.5 billion figure in context, the population of Los Angeles County (where Northridge is located) is 9,244,646, with 1,565,862 single-unit dwellings and 1,402,997 apartment units (as of January 1, 1995; California Department of Finance, 1995). This works out to insurance company payments of $1,352 for each man, woman, and child over a very large geographical area. The value of the total loss from the earthquake, including disaster assistance and uncompensated losses, is much greater.

The insured loss amounts were about one-third commercial and two-thirds residential. The high peak-ground acceleration of the earthquake was the primary cause of damage to building contents, chimneys, and

The first floor garage of this apartment building in Reseda, California, collapsed onto the cars parked inside during the Northridge earthquake of 1994 (USGS, J. Dewey).

garden walls. Building damage from landslide and liquefaction was not as common as in the Loma Prieta earthquake. While the insurance payments on this event exceed all of the earthquake insurance premiums collected in this century, this amount should only be compared to the current premium volume since so few structures were insured in past years, and so little premium collected.

A typical earthquake policy insures for loss against structural damage, damage to contents, and loss of use (residential) or business income (commercial). "Loss of use" covers the cost of hotel accommodations and meals until the structure is repaired, or it covers the loss of rental income on the house. "Business income" covers the loss of profits and the costs arising from shutting down the business (sometimes called "business interruption"). In the Northridge earthquake, for every $100 of insured residential damage, there was an average of $20 of content damage, and $10 of loss of use. It turned out that these ratios were the same for the 1989 Loma Prieta earthquake, even though the dollar amounts were much greater in Northridge.

The insurance industry had great difficulty in estimating what the ultimate losses would be from the Northridge earthquake. Property Claim Services (PCS) collects statistics on disasters and runs training classes on catastrophe claims handling for member insurance companies. PCS was the designated insurance industry collector of the estimates of the Northridge insured losses, as well as the media spokesperson for this catastrophe. It polled its members and announced estimates of the total industry insured losses periodically. Table 4-1, which provides a sum-

TABLE 4-1 Estimates of Insured Losses Caused by the Northridge Earthquake, January 17, 1994

Month of Estimate	Estimate of Total Losses (in billions of dollars)
February 1994	2.5
April 1994	4.5
June 1994	5.5
August 1994	7.2
October 1994	9.0
January 1995	10.4
March 1995	11.2
May 1995	11.7
July 1995	12.5

Source: Property Claim Services.

mary of these figures, shows that estimates of total insured losses grow over time as more accurate damage figures become available.

The California Insurance Department conducted its own surveys. The first one asked insurers to report their estimates as of October–November 1994, and the total reported was $8.8 billion. The survey conducted February–March 1995 reported that total losses were $10.6 billion, including loss adjustment expenses. Both of these surveys were consistent with the surveys in the above table. A short, last survey made in 1996 confirmed that the final figure is $12.5 billion for industry insured losses.

Earthquake damage losses are difficult to estimate, because the full extent of damage is not known until reconstruction and repair have been completed. This is why the initial estimates after the earthquake were about $2.5 billion and then grew monthly for more than two years. Table 4-2 shows the estimates of the final total losses paid or to be paid by insurers after the 1994 Northridge earthquake, according to the California Insurance Department survey of insurers conducted in February–March 1995.

As the table shows, insurance covers a wide variety of types of losses, and is an effective mechanism for compensating for earthquake damage. Even though the lines of insurance listed in the table are standard for reporting insurance premiums and losses, not every insurer reported losses the same way. For instance, it is likely that some insurers reported residential earthquake policy losses under the "homeowners' multiple peril" line. However, the distinction between residential and commercial should be accurate.

The number of automobile damage claims resulting from the Northridge earthquake was 32,249. News photos showed many cars crushed under apartment buildings, but most of the claims were for minor damage from falling objects. Many small losses were also reported under the regular homeowners' policy, whether or not the homeowner had an earthquake policy. Glass breakage and fire damage are covered under the homeowners' policy, and at a low deductible.

STATE GOVERNMENT INITIATIVES
FOR EARTHQUAKE INSURANCE

After every major earthquake, a flurry of legislation is proposed in the California legislature to deal with disaster recovery. Much of the legislation addresses the needs of people who do not have insurance or

TABLE 4-2 Northridge Earthquake Insured Losses, as Estimated by Insurance Companies as of February–March 1995

Line of Insurance	Number of Claims	EQ Losses Paid and Estimated	Loss Adjustment Expenses
Life	81	$1,985,913	$170
Accident and health	10,017	14,062,891	14,958
Fire, residential	424	12,238,044	569,571
Fire, commercial	211	141,602,269	2,216,404
Fire, undetermined	106	1,620,367	166,734
Allied lines (special)	3,278	118,653,759	5,750,356
Farm owners' multiple peril	537	3,039,707	207,604
Homeowners' multiple peril	74,471	929,312,456	40,511,945
Commercial multiple peril	9,374	1,096,236,983	26,796,625
Other liability	83	1,052,707	306,705
Misc. property, residential	3,570	40,291,899	465,906
Misc. property, commercial	1,321	528,630,245	11,273,250
Misc. property, undetermined	652	164,445,297	9,896,451
Earthquake, residential	185,180	5,521,488,790	186,477,890
Earthquake, commercial	3,939	1,059,013,892	33,354,708
Earthquake, undetermined	658	52,574,473	3,598,002
Workers' compensation	138	2,878,042	29,261
Automobile, personal	32,249	55,553,714	566,729
Automobile, commercial	846	1,716,603	39,854
Glass	295	523,695	5,102
Burglary and theft	3	4,837	0
Boiler machinery	17	5,135,114	19,122
Other, residential	5,041	43,742,677	2,117,551
Other, commercial	19	10,674,142	44,461
Other, undetermined	706	1,726,558	70,194
Late reported claims	405,253,516	57,897,448	
Totals	333,216	$10,213,347,588	$382,407,000

Note: Final losses and expenses are expected to be $12,500,000. However, the distribution of the losses among the lines should not change.

Source: California Insurance Department.

disaster assistance by providing for special tax deductions, postponement of payments, and lowering the assessed value of damaged homes for real estate taxes. The need to provide special relief has also spurred the state legislature to authorize special insurance funds to expand the number of people covered against future earthquake losses as well as the extent of their protection. This section summarizes two important pieces of legislation that addressed these issues.

The California Residential Earthquake Recovery Fund (1992)

At the time of the Loma Prieta earthquake in 1989, the standard earthquake insurance deductible was 10 percent (of the coverage on the structure). This meant that on a $150,000 house, the insured had to pay the first $15,000 of the cost of repairing the damage. After the Loma Prieta earthquake, the public protested to insurers and California legislators about such a high deductible. Many insurers responded by offering both 5 percent and 10 percent deductible policies. In the fall of 1991, the California legislature, after extensive public testimony, responded by establishing the California Residential Earthquake Recovery Fund (CRERF), which became effective January 1, 1992.

The sole intent of the CRERF program was to provide earthquake insurance to cover the 10 percent deductible. Regular earthquake policies remained unaffected. For about $60, the public could buy $15,000 of earthquake coverage (subject to a small deductible). The charge was included on the bill for the homeowners' insurance policy. It was a popular product (about 90 percent of homeowners bought it), and the price/coverage relationship was reasonable. The premiums were collected by insurers and forwarded to the CRERF and held by the state treasurer. This program was only in existence during 1992, after which it was repealed.

In 1992, there were 16 damage-causing earthquakes, including the Landers earthquake, which was one of the strongest ever recorded in California, although the damage was small because of its location. The CRERF paid all claims in full. Had the CRERF been in existence at the time of the 1994 Northridge earthquake, the expected loss covered by this policy would have been $1.2 billion (some estimates are higher). The fund would only have had about $400 million in funds to distribute, so it would have paid pro rata 33 cents on the dollar for the claims. This would have amounted to a payment of $500 to $800 to each homeowner with earthquake damage.

The CRERF program was repealed by the legislature at the request of the insurance commissioner for three reasons. First, the management and cost of the program became a financial burden, with over 70 people hired to manage the computer tracking of policies. Second, the earthquake claims were handled by independent adjusters, instead of the insurers writing the homeowners' policies. This turned out to be costly and inefficient. Third, and perhaps most important, the insurance commissioner was very concerned about the political problems that he thought might arise in the event that the CRERF could not pay its claims in full.

CRERF was an interesting program that was well received by the public, because it provided up to $15,000 in coverage for the deductible. This is not a significant amount of earthquake insurance protection on an individual home in California, and would only have provided meaningful protection if it supplemented a basic earthquake insurance policy covering the damage in excess of the underlying deductible.

The California Earthquake Authority (1996)

In 1985, the California legislature passed a law requiring insurers writing homeowners' insurance on one- to four-family units to offer earthquake coverage on these structures. There was no requirement that the owners had to buy earthquake insurance, only that the insurers had to offer it. This law was not instigated by consumers, but was passed at the urging of the insurance industry to overcome a recent lower court decision which had required greatly expanded coverage under the homeowners' policy, including earthquake damage. In other words, the lower court decision appeared to enable homeowners to collect for earthquake damage even though they had not purchased an earthquake endorsement, and the insurer had not received a premium for the earthquake coverage.

The lower court case was eventually overturned by the California Supreme Court, but the mandatory offer law remained in effect. Many people who were close to the drafting of this law believe that it was a poor solution to a legal problem that no longer existed. Furthermore, it took away from insurers the ability to manage their total earthquake exposure and forced them to insure structures which are so old or in such poor condition that they should not be provided with coverage.

The insurance companies lived with this law until the 1994 Northridge earthquake struck. The insured damage was beyond expectations, and when homeowners across the state heard about the average insurance payments of $30,000 to $50,000 after the 10 percent deductible for homes in the Northridge area, earthquake insurance became a desirable commodity in the minds of many homeowners. The insurance companies reevaluated their earthquake exposures up and down the state and decided that they could not risk selling any more earthquake policies. In view of the mandatory offer law, the only legal response they had was to stop offering new homeowners' policies, although in fact they did renew existing homeowners' policies even at the risk of the policyholder's opting to buy earthquake insurance.

In view of the desire of homeowners to maintain their earthquake coverage in the future, there was no possibility that the mandatory offer law could be repealed by the legislature. The California Insurance Department surveyed insurers and found out that up to 90 percent of them had either stopped selling new homeowners' policies or had placed restrictions on selling them. After extended discussions between the Insurance Department and the large insurers, an advisory group of insurers and actuaries proposed the formation of a state-run earthquake insurance company—the California Earthquake Authority (CEA).

The challenge from the start was to capitalize the CEA adequately. This challenge generated some innovative financing ideas. First, the department and the advisory group listed the possible sources of funding and chose these possibilities: (1) cash up front, (2) post-event assessments on insurers, (3) post-event assessments on CEA earthquake policyholders, (4) reinsurance, and (5) borrowing in the capital market.

Second, each of these possibilities was designated as a "block." Then the blocks were put into various patterns to decide which one made the most sense. In 1995, the state legislature passed a law, referred to as Assembly Bill 13, authorizing the insurance commissioner to undertake a feasibility study of the CEA proposal. The final block pattern, which was incorporated in this law, is shown in Table 4-3 and totals $10.5 billion in start-up funding.

The innovative features of this financing plan are the ability to pay for a large earthquake while committing relatively few dollars up front. At the time of the formation of the CEA, the capital market had never been used to back potential earthquake losses. In the proposed plan, capital markets were given the opportunity to cover the layer of losses from $7 to $8.5 billion. Before the markets had a chance to raise the full amount of capital, Warren Buffet of Berkshire Hathaway offered to cover the layer with his company's funds, and the insurance commissioner accepted his offer.

The excess-of-loss reinsurance commitment is $2 billion in excess of $4 billion. No reinsurance commitment this large has ever been made. A group of reinsurance brokers was chosen to determine if such a large placement was possible. Presentations were made to reinsurance companies in the United States and Europe. A long list of reinsurers agreed to commit small amounts which added up to $1.7 billion. The reinsurance that was placed was a two-year policy with an aggregate limit, a

TABLE 4-3 Capacity Participations in the California Earthquake Authority

Layer	Source of Funding	Total
up to $1 billion	INDUSTRY-CONTINGENT ASSESSMENT (to start the program)	$1 billion
$3 billion ($1–$4 billion)	INDUSTRY-CONTINGENT ASSESSMENT (after the earthquake)	$4 billion
$2 billion ($4–$6 billion)	REINSURANCE (no reinstatement)	$6 billion
$1 billion ($6-$7 billion)	POLICYHOLDER-CONTINGENT ASSESSMENT	$7 billion
$1.50 billion ($7–$8.5 billion)	CAPITAL MARKETS (bonds that default on interest)	$8.5 billion
$2 billion ($8.5–$10.5 billion)	INDUSTRY-CONTINGENT ASSESSMENT (after the earthquake)	$10.5 billion

provision that there would be no coverage after the limit had been exceeded, and a provision that the rate would be just under 15 percent rate on line (that is, 15 percent of the $1.7 billion, or $255 million, per year). This translates to very expensive reinsurance, but was the only choice, because the worldwide reinsurers are already heavily committed to reinsuring the commercial earthquake insurance market in California. The all-residential CEA program would have to be in addition to the reinsurers' commitment to the commercial earthquake insurance market in California.

The $10.5 billion total funding amount assumes that 100 percent of the current earthquake insurance policyholders will agree to buy coverage insured through the CEA, which would be limited coverage policies with a higher deductible (15 percent instead of 10 percent) and additional exclusions. If the CEA had been in business at the time of the Northridge earthquake, the CEA would have had to pay out about $4 billion. Therefore, the $10.5 billion figure was chosen so that the CEA would be able to pay for an earthquake two-and-a-half times the Northridge event.

When the CEA plan became operational at the end of 1996, insurers representing 72 percent of the homeowners' market elected to join the CEA program. All of the dollar amounts in Table 4-3 should be multiplied by 72 percent to get the actual layers for the program as it was implemented. For instance, the $2 billion reinsurance layer is now $1.44 billion, so the $1.7 billion in committed funds was more than adequate to cover this layer. The $1 billion first layer became $720 million, which was paid in by the participating insurers, and so on. At the time of the Northridge earthquake, there were approximately 1.9 million earthquake policies on single-unit dwellings. The CEA rate filing submitted to the California Insurance Department assumed that 72 percent times 1.9, or approximately 1.4 million policies, would be sold by the CEA.

The rates in many parts of California were set at higher levels than in the past, and the deductible was fixed at 15 percent. A 15 percent deductible is actually quite high. In order for a house to sustain damage to at least 15 percent of its value, the house must be located within 20 miles of the fault or on poor soil. The rates being charged for the policies with 15 percent deductibles have risen to about $3 per $1,000 of coverage for wood frame houses on good soil, and up to $6 or $7 per $1,000 for houses in higher-risk locations or near known faults. For a high-value house in a high-risk area, the premium can easily run into thousands of dollars per year.

From the consumer perspective, these increased rates and the unlikely chance of filing a claim due to the high deductible have led many policyholders not to renew their coverages. As of July 1997, the number of earthquake policies written through the CEA is below the level that the CEA promoters had anticipated. They anticipated writing about 26,600 policies per week (1,384,768 divided by 52). They are actually writing about half that many per week. Comments from insurance agents and the public indicate that the number of policies in some areas could be even less than half of the pre-CEA figures. Those who had earthquake insurance before and are now choosing *not* to buy the CEA policy have been complaining, as are those who are still renewing their coverages. This suggests that there is a strong interest among many property owners in protecting their homes against earthquake damage, but not by purchasing insurance.

Ironically, this concern could translate into renewed interest and investment in the retrofitting of existing homes, attention to building codes, and the future structural design of houses. Up until now California residents have shown a noticeable lack of interest in adopting mitigation measures. A 1989 survey of 3,500 homeowners in four California counties subject to the hazard reported that only between 5 and 9 percent of the respondents in each of these counties reported adopting any loss reduction measures (Palm et al., 1990). A follow-up survey by Palm and her colleagues in 1993 revealed that between 20 to 25 percent of the homes in the two counties affected by the 1989 Loma Prieta earthquake (Santa Clara and Contra Costa) had bolted their house to the foundation; less than 10 percent of homeowners in the two southern counties in the survey (Los Angeles and San Bernardino) had undertaken this measure (Palm, 1995).

The lack of interest in purchasing the CEA earthquake insurance raises a number of difficult political questions. The structure of the CEA funding arrangement was based on the assumption that most of the residents who had earthquake insurance would renew their policies, so that the CEA would receive sufficient premium income to cover the costs of those parties, providing coverage for each of the layers. With the unexpectedly large decrease in demand, it is unclear whether the CEA will have enough funds available to pay the reinsurers and Berkshire Hathaway.

The current lack of interest in this earthquake insurance means that even if there is a major earthquake in California next year with damage that exceeds the Northridge quake, it is unlikely that the insured losses would be as large, therefore it is most likely that the higher layers of reinsurance funding for the CEA will not be touched. This means that the reinsurers and Berkshire Hathaway have received premiums on coverage that is now almost risk-free. Whatever happens, the California Earthquake Authority has received worldwide attention, and much has been learned from the discussions and analytical work that have gone into this initiative.

THE CALIFORNIA EARTHQUAKE INSURANCE MARKET

The CEA was formed by the state legislature in order to relieve insurers of the risk of providing protection against earthquakes in compliance with the mandatory earthquake insurance law. The anticipation was that after the CEA plan was enacted, these insurers would then eagerly offer standard homeowners' insurance again. This has not happened: insurers are still being very cautious about selling homeowners' coverage.

The problem seems to be that these insurers are still liable for a major earthquake. As Table 4-3 shows, insurers that have elected to participate in the CEA will be subject to a post-earthquake assessment in the layer $3 billion excess of $1 billion. Such an assessment would be based on the insurer's market share of earthquake policies (including CEA policies). The only way an insurer can limit this potential assessment is to control its market share of earthquake policies. This can only be done by controlling the number of homeowner policies being sold. If an insurer sells a homeowners' policy, the insurer must offer to sell a CEA policy. The more CEA policies the insurer sells, the greater the potential assessment on that insurer from the CEA in the event of a major earthquake. Therefore, to reduce that potential assessment, the insurer must sell fewer CEA policies, but the only way to do that is to sell fewer homeowners' policies.

This section describes the private insurance market which offered earthquake insurance coverage to both homeowners and businesses prior to the formation of the CEA.

Size of the Market

Out of 800 property-casualty insurers active in California, about 175 insurers sold earthquake insurance prior to the establishment of the CEA in 1996. Most residential earthquake insurance was purchased from a few large insurers, such as State Farm, Allstate, Farmers, USAA (selling to the military), and the northern and southern California automobile clubs. Most commercial earthquake insurance was obtained through insurance brokers who assemble groups of insurers around the world to take a share of the large risks. The dollar value and premiums involved in insuring hospitals, factories, and office buildings can be quite large. Since most commercial earthquake insurance is sold in conjunction with other coverage that a business must have, it is difficult to know the exact amount of premium paid for the earthquake protection alone.

The property-casualty insurers in 1996 had a total premium income in California of $32.6 billion for all coverages, approximately $1.5 billion of which was for residential and commercial earthquake insurance. The figure for residential and commercial insurance was higher in 1996 than previous years, because rates for earthquake insurance were substantially increased after the Northridge earthquake.

Table 4-4 shows the amount of premiums paid for each major line of insurance in California for the year 1996. The table shows that the resources available to pay for another $12.5 billion in damages caused by

TABLE 4-4 Total Insurance Premiums for Policies Written in California, 1996

Line of Insurance	Premiums
Homeowners' Multiperil	$ 3,027,060,068
Commercial Multiperil	3,035,924,265
Workers' Compensation	4,935,028,301
General Liability	2,168,035,307
Private Auto Liability	7,688,676,280
Commercial Auto Liability	1,333,483,182
Private Auto Physical Damage	4,409,374,145
Commercial Auto Physical Damage	474,574,965
Earthquakes	1,500,000,000[a]
All Others	4,037,524,810
Total	$32,609,681,323

[a]Estimated.

the Northridge earthquake are quite limited, since each of these lines of insurance has its own losses and expenses. Any and all of these lines of insurance can also have claims arising from an earthquake. The insurance industry's invested reserves and surplus are intended to support the continuing business (such as automobile and workers' compensation), not solely to pay for catastrophic events. The Northridge earthquake was clearly subsidized by the other lines of insurance and the business written in other states.

In general, only large multiline, multistate companies insure against catastrophes. (A multiline, multistate company is an insurance company licensed in many states to sell all types of property-casualty insurance, such as workers' compensation, automobile, fire, earthquake, and commercial coverage.) The small insurance companies operating only in California do not have the financial resources to pay for a large earthquake; many small insurers barely survived the Northridge earthquake. The problem is that there are not enough large insurers around, and the cost of natural catastrophes is rising rapidly.

Many insurers are mutual insurers, which are owned by the policyholders, not stockholders. The central problem for mutual insurers is that in the event of a large earthquake that depletes their capital, they have nowhere to go to raise new capital. They can only raise insurance rates. Stock insurers have the advantage that they can sell more stock to raise additional capital after an earthquake. After a major earthquake, a somewhat perverse situation often exists in which the stock market will increase the value of an insurer's stock price in the anticipation that the insurer will be able to raise future earthquake insurance premiums and increase their profits. So it has not been that difficult for stock insurers to raise additional funds after an earthquake.

Number of Earthquake Insurance Policies

In connection with several proposals affecting earthquake insurance in the 1995–1996 legislative session, the California Insurance Department issued a special questionnaire in 1995 to all insurers writing earthquake insurance in California asking for detailed information on homeowners' policies and earthquake insurance policies in force at the end of 1994.

Table 4-5 shows the number of residential policies in force in California at the end of 1994 based on that survey. The table includes condominium and renter policies, but not mobile home policies. Homeowners' (HO) policies are specifically designed for certain types of housing units. The standard homeowners' policy is an HO-3. The policy for apartment renters is an HO-4; for condominium owners it is an HO-6. Earthquake coverage (EQ) appears as an endorsement to the respective policy. At the time of the 1994 Northridge earthquake (Los Angeles County) about 40 percent of the homes in the Northridge area had earthquake insurance. This percentage did not change over 1994, because insurers tried not to increase the number of earthquake policies during that year.

After this survey was taken, the Department of Insurance approved substantial earthquake rate increases, and an increase in the deductible for many insurers from the 10 percent that was common at the time of the Northridge earthquake to 15 percent of structure coverage. The bottom of the table shows the number of insurance policies as a percent of the number of insured and uninsured units in the state. The number of renters with insurance is quite low, as might be expected.

THE BUSINESS OF EARTHQUAKE INSURANCE

Insurance is a business: it involves accountants, actuaries, salespersons, claims adjusters, managers, executives, and so on. Insurance is a product: it focuses on a market; it has a value to the customer; and it has a price (or premium). But insurance has one feature that distinguishes it from most other consumer products: the cost of the product to the in-

TABLE 4-5 Percent of Dwellings, Condos, and Apartments with Earthquake Insurance as of the end of 1994

County	Dwellings Number HO[a]	Number EQ[b]	% EQ[c]	Condos Number HO	Number EQ	% EQ	Renters Number HO	Number EQ	% EQ
Alameda	285,949	123,418	43.2	17,637	9,234	52.4	29,983	15,636	52.1
Contra Costa	215,930	75,114	34.8	21,636	9,286	42.9	19,374	8,306	42.9
Fresno	138,478	18,715	13.5	3,045	617	20.3	10,728	1,655	15.4
Kern	125,744	28,571	22.7	2,329	613	26.3	6,706	2,374	35.4
Los Angeles	1,575,670	573,408	36.4	104,352	62,119	59.5	109,209	62,119	56.9
Orange	459,346	173,634	37.8	80,559	38,392	47.7	50,033	26,839	53.6
Sacramento	278,466	23,881	8.6	8,357	1,420	17.0	24,808	5,277	21.3
San Diego	490,400	122,133	24.9	56,867	23,576	41.5	60,299	28,101	46.6
San Francisco	117,161	41,634	35.5	10,278	5,310	51.7	32,691	20,536	62.8
San Mateo	150,109	68,565	45.7	11,731	6,267	53.4	20,124	10,548	52.4
Santa Clara	323,866	150,551	46.5	25,524	13,722	53.8	38,945	20,231	51.9
Santa Cruz	57,161	24,118	42.2	3,167	1,537	48.5	5,201	2,464	47.4
Rest of State	2,724,270	724,954	26.6	130,864	59,991	45.8	201,190	76,361	38.0
Totals	6,302,093	1,931,449	30.6	434,028	212,947	49.1	549,206	254,850	46.6
Total Insured + Uninsured	6,481,927			833,118			3,865,777		
Percent Insured	97.2		29.8	52.1		25.6	14.2		6.6

[a]Number HO = Number of HO Policies
[b]Number EQ = Number with EQ Coverage
[c]%EQ = Percentage with Earthquake Coverage

Source: California Insurance Department, California Statistical Abstract (California Department of Finance, 1995).

surer is determined only *after* the product is sold, because the cost depends on claims paid out during the policy period. Therefore, the expected losses and other expenses must be estimated beforehand.

Estimating losses and expenses is the work of actuaries, who, as an example, project losses due to automobile accidents based on past accident history, inflation in medical costs, and changes in the tort laws. Similarly, life insurance actuaries project insurance payments based on morbidity and mortality tables, which were derived from the historical experience of people who had life and health insurance in the past. The insuring of natural disasters presents the ultimate challenge to actuaries because the past cannot necessarily be used to project the future. Instead, actuaries rely more and more on scientific and engineering knowledge when trying to quantify these low-frequency, high-severity events.

Rating and Underwriting Earthquake Risks

Insurers manage the business of insurance through two processes: rating and underwriting. *Rating* is the process of determining the proper amount to charge per $1,000 of coverage, given the risk characteristics of the peril being covered. In the case of earthquake insurance, the rate could be based on the susceptibility of the structure and contents to shake damage, the proximity to known faults, the characteristics of the faults, and the soil conditions under the structure.

Prior to the mid 1980s, insurers charged one rate for earthquake insurance (about $2 per $1,000 of coverage and a 5 percent deductible) for a standard house anywhere along the coast. They charged a lower rate for the interior of the state and a higher rate for masonry homes. So few homes had earthquake insurance that the insurance industry did not pay much attention to the actuarial correctness of these rates.

In more recent years and prior to the passage of the CEA, earthquake rates were as complicated as automobile rates (which vary by territory, type of car, usage of car, driver's age, sex, marital status, and driving record). When earthquake rates are determined by actuaries, the actuaries normally do not include the potential losses from the mega-catastrophe earthquake or even the largest earthquake that occurs once every two or three hundred years or more. The rates are usually based on all of the small, medium, and large earthquakes that are likely to occur only in the next one hundred years. In addition, a guide often used by insurers is to ask how many years of premium it would take to pay for a large earthquake. Many insurers regard 5 to 10 years of premium to pay for a large earthquake as a reasonable level of risk to take.

Underwriting is the process of determining at what price to insure the risk. If the underwriter believes that the price should be higher than the rate allowed by the regulator or the rate the insured would be willing to pay, then the underwriter would decide not to offer coverage for a particular risk. For example, the underwriting process might result in a decision not to insure a structure against earthquakes in the case of very poor construction, very poor soil conditions, or close proximity to major faults. In the case of a house on the side of a hill, for instance, which could slide in even a small earthquake, a claim on the insurance policy is likely to be made and the amount of the claim is very uncertain, so the risk could not be priced reasonably and would be rejected.

Underwriting can be used as a powerful mitigation tool by requiring structures to be retrofitted in order to become insurable. Underwriting may conclude that deductibles are necessary to eliminate the numerous small losses that occur because of an earthquake. Finally, underwriting can indicate the need for exclusions to eliminate claims for swimming pools, decking, brick veneer, and non-seismic earth movement such as landslides.

Estimating Damages

In order to undertake rating and underwriting, insurers need information from engineering and scientific studies. Some of the key concepts that are relevant in this regard are damage ratios and exceedance probabilities.

A *damage ratio* is the ratio of the expected insured loss to the replacement value of the structure and contents, assuming that the building was on good soil and is an average value for its area. Soil maps and fault maps must be expressed in terms of some intensity scale that can be applied to structures with known damage ratios to get an estimate of the expected average annual loss to that structure at that location. Insurers also need to know the frequency distribution of damage ratios for each zip code; such a distribution would give insurers the number of small losses as well as large losses, information that is needed to determine the effect of changing deductibles.

Insurers and reinsurers also need to know what are called *exceedance probabilities*, such as the probability of exceeding $4 billion in insured losses from a given earthquake event. When the California Earthquake

Authority was proposed to the legislature, all of this information was estimated. At present, these estimates are based on insurance claims statistics from past earthquakes for certain types of structures, from government research studies (and maps), and from earthquake computer models.

While it may be very difficult to predict when and where the next earthquake will occur, it is important to know how the amount of damage varies with the distance to the epicenter and the magnitude of the earthquake. It is very important and useful to the insurance industry to know the *relative* amount of damage that will occur for different types of building construction and for different types of soil conditions. The relativity between the factors is important because this is reflected in the proper premium rate to charge for each building at a particular location. For example, if the damage to a masonry building is expected to be ten times greater than the damage to a wood frame building, then the loss portion of the premium rate should also be ten times greater. In addition, the relativity between the factors affects how an insurance company manages its portfolio of risks by enabling it to reduce the high concentrations of risks in certain areas.

The last major published study undertaken by structural engineers to estimate damage ratios was ATC-13, published by the Applied Technology Council in 1985. In view of the extensive building damage experienced during the 1994 Northridge earthquake, the insurance industry would welcome a confirmation or update of that study by the structural engineering profession. The ATC-13 study (funded by FEMA) is the basic analysis used by the commercial earthquake modelers, and therefore the basic analysis used by the insurance industry. The California Insurance Department and the U.S. Geological Survey funded two studies on damage ratios based on insurance industry claims statistics. (See Steinbrugge and Algermissen, 1990, on insured losses before Loma Prieta; and Steinbrugge and Roth, 1994, on insured losses from Loma Prieta.)

Insurers need to work closely with earth scientists and engineers in interpreting available data. For example, damage patterns may be fairly predictable with a strike/slip fault (as distance from the fault rupture plane increases, damage usually decreases, with the exception of poorly constructed buildings on poor soil). However, damage patterns may be more complicated with a thrust fault, because the fault plane dips at an angle; serious damage may occur in places that might not be expected using traditional models that work with surface fault traces. In the special case of blind thrust faults, damage patterns may be even more difficult to predict. By working closely with scientists and engineers, insurers can make decisions that reflect a sophisticated understanding of the earth-quake risk.

Estimating the Probability of Earthquakes

Another area of scientific knowledge that is valuable to insurers is the study of earthquake probability. Over the years, seismologists have been estimating the likelihood of a given fault rupturing again. These estimates are based partly on the length of time that has passed since the last rupture and on the long-term slip rate. However, a fault can also rupture when nearby faults rupture. The most recent study by the Southern California Earthquake Center (Working Group on California Earthquake Probabilities, 1995) presented a consensus estimate of the probability of earthquakes of specific sizes for 65 seismic zones in southern California. A working group of leading seismologists and geologists from the California state government and its universities and from the U.S. Geological Survey made up the consensus team. A second report, prompted by the 1992 Landers earthquake, evaluated the probability of a major earthquake within the next 30 years on each of the major known faults in southern California. The report concluded that there is a high probability of significant seismic activity in the next 30 years in the Santa Barbara, Palmdale, and San Bernardino areas of southern California.

The conclusions of the USGS study are that the occurrence of an earthquake in southern California of a magnitude (M) on the Richter scale of 6.0 or greater is expected each 1.6 years, on the average, during the next 30 years. (This recurrence interval is about double the observed rate since 1850.) The probability of a magnitude of 7.0 or greater in the next 30 years is estimated at 86 percent. (The predicted rate of earthquakes for M of 7.0 or greater is also about double the rate since 1850.) Great earthquakes of a magnitude of 7.8 or greater have an estimated 6 percent to 9 percent probability of occurrence in the next 30 years. This is particularly ominous since, depending on the location of the epicenter, the effect of such an earthquake on a densely populated area would be so devastating that it could be classified as a mega-disaster.

Role of the California Insurance Department

Before the CEA was established, California state regulations required all earthquake insurance rates to be filed with and approved by the California Insurance Department (CID). Concern about the exposure of the insurance industry to earthquakes greatly increased in the aftermath of the San Fernando earthquake in 1971, leading the California Insurance Department to issue Ruling 226. This ruling required all licensed insurers to report yearly on their insured exposures for earthquake shake damage on residential and commercial structures in the state, via a detailed questionnaire. At that time, the percentage of homes and commercial structures insured for earthquake damage was less than 10 percent, and the insurance losses from the San Fernando earthquake were about $46 million. Since then, the demand for earthquake insurance has grown substantially, along with a large increase in housing and commercial building values.

These annual earthquake questionnaires also elicit information used for estimating the probable maximum loss, or expected insured loss for structural and contents damage from a major earthquake in each designated earthquake zone. From these questionnaires, the insurance department compiles estimates of the aggregate insured PML losses by earthquake zone and publishes these results. The state is divided into eight zones, but Zones A (San Francisco) and B (Los Angeles/Orange County) are the most important. The questionnaires for individual insurers are not made public.

The California Insurance Department uses these questionnaires to monitor the amount of earthquake exposure of each insurance company in relation to its financial strength. The questionnaire shows the amount of earthquake insurance by earthquake zone and by residential and commercial building classes. In addition, there is detailed information on reinsurance. This questionnaire was an impetus to the development of the many earthquake simulation models that are now available. It has also encouraged insurers to improve their knowledge of seismicity, geology, and structural engineering, and to develop their own in-house expertise.

The California Insurance Department also, under its regulatory authority, conducts detailed surveys of insurance industry losses after major earthquakes. Some of the statistics from these surveys are shown in the tables in this chapter, and some research publications based on these surveys are listed in the references to this chapter (Roth, 1995-1996; Steinbrugge and Algermissen, 1990; Steinbrugge and Roth, 1994).

MARKETING PRIVATE EARTHQUAKE INSURANCE

Building on the conditions of insurability detailed in Chapter 2, this section examines whether earthquake insurance can be marketed by private companies at a price that reflects the risk and reduces the chance of insurer insolvency to an acceptable level, while still satisfying the regulator, the California Department of Insurance.

Probable Maximum Loss

In Chapter 2, we indicated that insurers are concerned with the probable maximum loss that they can suffer from a single disaster. The following example illustrates how PML is calculated and used in relation to earthquakes. An insurance company has sold earthquake insurance on 100,000 homes in the San Francisco Bay Area. Since their average replacement cost is $200,000, the aggregate replacement cost of the homes might be $20 billion ($200,000 × 100,000). It is highly unlikely that any event, even one like the 1992 Oakland fire, could destroy all 100,000 homes. In the event of an earthquake, most of the homes would be only partially damaged, and many would not be damaged at all.

Now suppose the California Insurance Department wants to determine the PML for single-family residences in conjunction with a 10 percent deductible earthquake policy for insurers who are selling policies in the San Francisco Bay Area earthquake zone. The CID looks at damage statistics compiled by structural engineers who inspected typical houses damaged in past California earthquakes, and then at estimates derived from these statistics of what the damage would be to these houses if the earthquake had been a major one. Using this information, the CID indicates to insurers that the PML percentage factor is 1.7 percent of the insured value. This means that the expected loss to the insurance company from a major earthquake based on the average PML for the above example of 100,000 homes with coverage against earthquakes would be $340 million (1.7 percent × $20 billion).

PML estimates are sometimes based on the largest expected earthquake in the next one hundred years, and sometimes they are said to be based on an earthquake at the 90th percentile of damaging earthquakes in size. *The probable maximum loss estimate does not include any estimate of damage from a mega-catastrophe earthquake, nor does it include the possibility of an unusually large number of small or medium-size earthquakes.*

There are different PML factors for different deductibles, and for commercial buildings of differing construction. The goal is to estimate what would happen if the maximum likely fault rupture occurred at one of the large faults in the affected zone. The PML percentage varies from fault zone to fault zone (Roth, 1995-1996).

From an insurance company underwriter's point of view, $340 million is a much more useful number than the $20 billion figure. If the insurer decides to commit $250 million of the insurance company's net worth to a possible earthquake event in the San Francisco Bay Area earthquake zone, the underwriter will conclude that there are too many earthquake policies in that zone.

An increasing number of insurance companies now prefer to use *exceedance probability* (EP) curves. An EP curve specifies, for a given year, the likelihood of an earthquake that causes damage equal to or greater than "X" dollars. The value of "X" is varied over the entire spectrum of potential damage from earthquakes. In other words, a PML is just one of many values used in constructing an EP curve. This curve is normally generated by a computer model.

Capacity

As pointed out in Chapter 2, there is a direct relationship between the PML and the maximum amount the insurer is willing to cover in one disaster zone, which is called its capacity. After the 1994 Northridge earthquake, insurance managers reevaluated their PML exposures in relation to their chosen capacity for California earthquake exposure. Many insurers announced that they would not be selling any more earthquake insurance policies, but would be renewing existing policies. Of course, the Northridge earthquake showed many property owners what an earthquake can do to a home. When these homeowners decided that they wanted earthquake insurance, many found that it was not available.

An essential part of controlling capacity is the use of reinsurance. About half of earthquake reinsurance is sold by reinsurers in the United States and the other half by reinsurers around the world, mostly in Europe. Although the concept of reinsurance is simple, contracts can get quite complicated. Practically every commercial earthquake policy is reinsured with other insurers, usually on a building-by-building basis, or what is called "per risk" insurance. Very few residential earthquake policies are reinsured individually since the expected damage per house is relatively small; earthquake insurance losses on homes are usually aggregated, and the reinsurance policy prevents the aggregate loss from getting too high.

Insurability Issues

Insuring earthquakes in a financially responsible manner is essentially a matter of limiting the potential insured loss in each location where such an event is likely to occur. It also means assuming that no mega-earthquake occurs during the period of coverage (that is, an earthquake of magnitude 7.8 or greater). The worst situation for an insurer is to have all of its earthquake policyholders concentrated in one area, in fully insured high-valued houses, near a fault where soil conditions are poor (on hillsides or with a high probability of liquefaction). Under these conditions, the PML from one event would be very high indeed.

It is often said that the purchase of earthquake insurance must be mandatory in order to spread the risk and make earthquakes insurable. Such a provision has been attached to many governmental disaster programs and proposals. *There is no actuarial or scientific basis for making the purchase of earthquake insurance mandatory in California or anywhere else.* Over two million residences, out of six million, are insured against earthquakes throughout California, all policies voluntarily pur-

---

SIDEBAR 4-1 ▪▬▬▬▬▬▬▬▬▬▬▬▬▬▬▬▬▬▬▬▬▬▬▬▬▬▬▬▬▬▬▬▬▬▬▬▬▬▬

Making Earthquakes More Insurable

In order to make the potential loss from an earthquake more insurable, the strategy is essentially one of divide and conquer and involves these essential points:

- Spread the risks geographically, to reduce the proportion of total risks insured that can be affected by one earthquake. This is done by identifying the known fault areas and spreading the number of insured risks among these areas.
- Control the amount of loss that is probable from each risk. This is done by imposing large deductibles, by not insuring high-valued dwellings, by imposing exclusions (such as brick veneer, swimming pools, garden walls), by limiting the coverage on contents, and by requiring earthquake retrofitting to lessen the building's susceptibility to earthquake damage.
- Make the rates reflect the risk of loss. This is to avoid underpricing earthquake insurance in high-risk areas, which would attract a large number of policyholders, with adverse financial results for the insurer in the long run.

chased. On an actuarial basis, whatever can be gained by diversifying geographically was gained long ago.

If the political objective is to make owners of low-risk buildings subsidize owners of high-risk buildings (such as those built of unreinforced masonry in fault areas), then it may be necessary to impose a mandatory (tax) scheme. Furthermore, it is not clear why earthquake insurance should even be sold for many old buildings, because earthquake policies are usually replacement coverage policies which pay the cost of replacing destroyed buildings with brand new ones, often of far greater value than the old.

As a final argument against making the purchase of earthquake insurance mandatory, the strategies listed above to increase insurability would not be possible since there would be no opportunity for the insurer to select risks. Therefore, making the purchase of earthquake insurance mandatory would actually reduce the insurability of earthquakes for the insurer or the government pool providing the coverage. In California, the law requiring insurers to offer earthquake insurance to everyone who buys homeowners' insurance has caused a serious restriction in

---

- Judiciously purchase reinsurance. Reinsurance can be purchased to limit the amount of loss on each risk, or on the aggregate of all risks combined. Since reinsurance is expensive, the proper coverage and price must be worked out carefully. On commercial risks, reinsurance is a necessary element of the strategy.
- Utilize the vast available research in seismology, geology, and structural engineering. An effective strategy to improve the insurability of earthquakes must include the exploitation of this available research. In the Loma Prieta earthquake, the damage-causing liquefaction areas in the Marina district and the hillside areas in Santa Cruz were all known.

These measures add up to a new form of insurance management which is much more scientific, financially and actuarially sound than has been seen in the insurance industry before. Using this form of management, it is possible to make small and medium-size earthquakes insurable; it is not useful with the very large earthquakes, where the insurance industry's own survival is at stake.

the availability of earthquake insurance for this reason and led to the formation of the CEA. Although the homeowners are not required to accept the offer, the insurers are required to make earthquake insurance available and thus cannot select risks to control the PML level by territory. This prevents insurers from increasing the insurability of their portfolio of earthquake risks.

PROPOSALS FOR FEDERAL GOVERNMENT EARTHQUAKE INSURANCE

Since the 1980s, the insurance industry has been advancing proposals to the U.S. Congress to establish a federal natural hazards insurance program, somewhat like the National Flood Insurance Program (NFIP). First addressing the earthquake peril, these proposals have been expanded to cover most natural hazards in order to gain broader support in Congress. These proposals have been motivated by the following factors:

- The risks involved in insuring natural disasters are so uncertain and so large that most insurers do not want them to be part of the costs of being in the insurance business. Insurers are business enterprises which must attract capital and make profits in competition with other businesses.
- Government programs can accumulate premiums and investment income free of taxes and therefore can accumulate funds at a much greater rate than insurers can.
- A single pot of money is sometimes assumed to be the best "spread of risk."
- A government-run insurance program could reduce the heavy governmental reliance on post-event disaster relief loan and grant programs.
- Efforts to mitigate earthquake damage may be more effective if tied to a government insurance program (as is the case with the NFIP).
- The NFIP has been held up as a model federal catastrophe insurance program.

As reasonable as these factors seem to be, they are in conflict with some sociological, economic, and actuarial teachings and principles. These teachings and principles focus on the "true" cost of such government programs (the opportunity cost of the funds), the foregone benefits of a competitive insurance marketplace (e.g., cost efficiency and rate competition), and the absence of consumer choice (the ability to decide whether to purchase coverage). In simple and political terms, the controversy boils down to determining what government can do best, what the private insurance industry can do best, and what the fairest allocation of scarce government resources is among all of the competing demands on government budgets and commitments.

California and other western states would be the primary beneficiaries of the earthquake portion of this program, but there are states with significant exposure to wind and hurricane that would benefit as well. In recent years, it has been easy to find members of Congress willing to sponsor legislation to establish such a program, but actual forward movement of these bills has been slow.

This effort to push for federal legislation has actually been very fruitful in promoting and funding research studies on earthquake insurance issues. Over the years, the major federal agencies have been sponsoring research studies in earthquake science as well as actuarial science, economics, finance, and other social sciences. These studies have greatly increased our knowledge and understanding of the scientific and financial aspects of earthquakes and natural hazards in general.

Advancements in the science of earthquakes are needed to predict the probability and magnitude of future earthquake events. Research programs in economics and actuarial science are needed to plan for the necessary funding for earthquake recovery, and to devise the optimum allocation of resources after the event in order to promote speedy economic recovery of the affected region and the rebuilding of the damaged residential, commercial, and public structures.

The reluctance of Congress to act seems to stem from a lack of consensus on what will work, a realization that the problem is not well understood, and a concern about adequate provisions for mitigation. The nation has come a long way in understanding the science and economics of earthquakes, and more can be learned and understood. This is an international issue. China, Japan, New Zealand, and other countries are working as hard as the United States to find a workable solution to responding to natural disasters.

Rationale for a Federal Reinsurance Program

There is no question that in past natural disasters, the government (state or federal) has been the primary source of relief and economic recovery. However, as in the case of medical care and welfare, it is not easy to determine the best allocation of responsibilities between the federal and state governments and all other stakeholders.

The proposal to establish a federal earthquake insurance program was originally based on the premise that these disasters were uninsurable and that the solvency of many, if not most, insurers was at stake, because of the huge potential insured losses that a large earthquake can cause. Today, insurers are more sophisticated and much better able to estimate probable maximum loss values and capacity limits, so that solvency and insurability are better managed, but they are still very real concerns.

As an example, the California Insurance Department did not have to take over any insurers because of excessive insured losses after the Northridge earthquake, which was a moderate-size earthquake. However, the insurers did take immediate steps to control the PML levels by restricting the issuance of new earthquake policies. This led to the restricted availability of earthquake insurance and the formation of the CEA.

As an alternative to the CEA, the federal government *reinsurance* program would provide the insurance industry with additional capacity to market earthquake coverage. The main problem that the insurance industry has is its vulnerability to the mega-catastrophe event, an earthquake so large and devastating that whole communities are practically destroyed over a wide geographical area, leaving some insurers insolvent, and seriously depleting the capital base of most other insurers. The 1964 Alaskan earthquake was an event of great magnitude over a wide geographical area. Although the area was sparsely populated, only federal aid and the discovery of oil enabled the region to recover economically. Had this mega-catastrophe event occurred in Los Angeles in 1995, the federal government would have had to consider federal loans or grants to the insurance industry!

SUMMARY AND CONCLUSIONS

The public demand for earthquake insurance has grown dramatically in recent years and, consequently, so has the insurance industry's exposure to large amounts of insured earthquake losses. Ordinarily insurance deals with situations that involve many small losses, the total of which is predictable based on past loss experience. Earthquake insurance, on the other hand, deals with low-frequency, high-severity events where past history is not useful for predicting the future loss experience at a particular location. Instead, the insurance industry must rely on engineering, geological, and seismological information and expertise to make estimates of the potential loss exposure to a group of insured buildings in a particular fault zone.

An insurer uses this information to manage its portfolio of earthquake risks to make sure that the insurer's potential loss does not exceed the insurer's capacity to pay the losses. This control of earthquake exposures means that earthquake insurance will not be available to all homeowners and building owners who want it. In California, the California Earthquake Authority is a government-sponsored attempt to make residential earthquake insurance available where the insurance industry has not been able to provide it.

There have been several damage-causing earthquakes in California recently and there will be more. The insurance industry, the government, and the scientific community need to be in a partnership to establish the most appropriate coordinated program for financing the essential economic recovery after an earthquake.

CHAPTER FIVE

Hurricane Insurance Protection in Florida

EUGENE LECOMTE AND KAREN GAHAGAN[1]

A S POINTED OUT IN CHAPTER 2, windstorm, which includes tornadoes, is covered as a peril in the residential and commercial property insurance policies in use today. This standard feature provides for the direct physical loss of or damage to covered property caused by or resulting from windstorm.

The passage of time and the regularity of windstorm events, hurricanes, and winter storms have confirmed the importance of this insurance. The recent spate of hurricanes to strike the state of Florida, Hurricane Andrew in particular, have revealed the extent to which this coverage is a necessity, as well as the precarious financial position in which many Florida insurers now find themselves. Today there is a question as to whether the voluntary insurance market can provide affordable coverage to customers who seek it and still ensure the long-term solvency of firms in the industry.

---

[1]Ron Bartlett, Clem Dwyer, Mark Johnson, Elliott Mittler, and Jack E. Nicholson provided contributions to this chapter.

FLORIDA'S VULNERABILITY TO HURRICANES

Florida is the state geographically and historically most vulnerable to hurricanes. North Atlantic hurricanes normally form out of tropical depressions off the coast of West Africa and move in a west-to-northwest direction through the Atlantic and Caribbean, frequently toward Florida's coast. They are also known to form in the Gulf of Mexico, the Caribbean, and off the coast of the United States. The North Atlantic hurricane season occurs during the months of June through November, with September generally having the largest number of hurricanes.

From 1871 to 1993, nearly 1,000 tropical storms of varying intensity occurred in the North Atlantic, the Caribbean Sea, and the Gulf of Mexico. Of these, about 180 reached Florida, with 75 known to have had hurricane-force winds (wind speed of 74 mph or greater) and 105, tropical-storm-force winds (39–73mph) (Doehring et al., 1994). During this century, the occurrence of tropical storms has not been random but has exhibited a cyclical tendency, with periods of activity and inactivity lasting two or three decades or longer. For example, the period from the late 1940s through the late 1960s had a much larger number of hurricanes (i.e., a strong cycle) than did the 1970s and 1980s (i.e., a weak cycle). Also, between the years 1900 and 1992 Florida had been hit directly by a little more than one-third of the hurricanes that struck the United States. This is far more than that experienced by any other states on the Atlantic or Gulf coastal states, as indicated in Table 5-1.

TABLE 5-1   Number of Hurricanes Striking Florida and the United States Mainland, 1900–1997

Area	Damage Potential Category[a]					Total	All Major Hurricanes[b]
	1	2	3	4	5		
Florida	17	16	17	6	1	57	24
Total U.S. (Texas to Maine)	59	36	47	15	2	159	64

[a]From the Saffir/Simpson Hurricane Damage Potential Scale. *Category 1*: 74–95 mph (minimal); *Category 2*: 96–110 mph (moderate); *Category 3*: 111–130 mph (extensive); *Category 4*: 131–155 mph (extreme); *Category 5*: more than 155 mph (catastrophic).
[b]Greater than or equal to Saffir/Simpson Category 3.

Source: NOAA, Tropical Prediction Center.

Recent Hurricane Activity

After Hurricane Andrew in 1992, which caused insured losses of $15.5 billion, Florida was spared hurricane activity in 1993. In the 1994 season, Tropical Storm Alberto caused an estimated $95 million in insured losses, and Hurricane Gordon caused approximately $60 million in insured losses.

The year 1995 was recorded as one of the most active years of this century for storms. The official count for named storms in 1995 was 19, and there were 115 named-storm days (Gray, 1995). Two 1995 hurricanes, Erin and Opal, caused significant damage to coastal property in Florida. Current estimates of insured losses for Erin are $375 million, and for Opal, $2.1 billion, making Opal the third most costly insured hurricane in U.S. history, after Andrew and Hugo.[2] Fortunately, losses from Opal did not exacerbate adverse conditions existing in the insurance marketplace in Florida; most of Opal's damage was caused by storm surge, coverage for which is provided by the National Flood Insurance Program.

It has been suggested that the exceptionally active 1995 hurricane season is an indicator of a move toward a new, strong cycle. William M. Gray, of the Department of Atmospheric Sciences at Colorado State University, publishes annual forecasts and periodic updates of Atlantic tropical cyclone activity. He believes the United States may be on the verge of an active hurricane cycle that could last anywhere from 20 to 50 years. This would mean that intense, damaging storms would threaten once every two years instead of the current rate of about once every seven years (Vowinkel, 1995).

Complicating the task of predicting hurricane activity is the growing realization that weather in the southeastern United States may be influenced by the El Niño Southern Oscillation (ENSO) phenomenon. According to O'Brien et al., Atlantic-based hurricanes are more likely to make landfall in a regular year rather than an El Niño year; specifically, "it is 2.2 times more likely to have two or more land-falling hurricanes on the United States" if El Niño has not occurred the previous winter (O'Brien et al., 1996).

---

[2]The source for estimated insured losses is Property Claim Services (PCS). PCS assigns a catastrophe serial number when the insured loss from an event is expected to be at least $5 million to property insured under fire and allied lines coverage, inland marine fixed property coverage, and comprehensive automobile coverage. The occurrence must also result in a significant number of claims.

Increasing Insurance Exposure

Each year approximately 130,000 new households are established in Florida, drawn by the attractive climate and vigorous economy. These demographic changes play a significant role in increasing Florida's vulnerability to hurricanes. The potential for loss of life and destruction of property continues to expand as the state's population grows and its structures multiply. It is not surprising that five of the ten most costly hurricanes in terms of insured losses have occurred in Florida, as shown in Table 5-2.

Florida's coastal county population rose from 7.7 million to 10.5 million between 1980 and 1993, an increase of 37 percent. As shown in Table 5-3, over three-fourths (78 percent) of Florida's population resides in counties that are adjacent to the Gulf or Atlantic Coasts (IIPLR, 1995).

TABLE 5-2   The 10 Most Costly Insured Hurricanes

Dates	Place	Hurricane	Estimated Insured Loss
1992, Aug. 23–24, 25–26	FL, LA, MS	Andrew	$15,500,000,000
1989, Sept. 17–18, 21–22	U.S. Virgin Islands, PR, GA, SC, NC, VA	Hugo	4,200,000,000
1995, Oct. 4–5	FL, AL, GA, SC NC, TN	Opal	2,100,000,000
1992, Sept. 11	Kauai and Oahu, HI	Iniki	1,600,000,000
1979, Sept. 12–14	MS, AL, FL, LA, TN, KY, WV, OH, PA, NY	Frederic	752,510,000
1983, Aug. 17–20	TX	Alicia	675,520,000
1991, Aug. 18–20	NC, NY, CT, RI, MA, ME	Bob	620,000,000
1985, Aug. 30–Sept. 3	FL, AL, MS, LA	Elena	543,300,000
1965, Sept. 7–10	FL, AL, MS	Betsy	515,000,000
1985, Sept. 26–27	NC, VA, MD, DE, PA, NJ, NY, CT, RI, MA, NH, VT, ME, TX	Gloria	418,750,000

Source: Insurance Information Institute from estimates provided prior to 1984 by the American Insurance Association; thereafter, by the Property Claim Services division of the American Insurance Services Group, Inc.

TABLE 5-3 Population Distribution in Florida, 1980–1993

Location	1980 Population		1993 Population		Percent Change 1980–1993		1993 Coastal Population as % of Total
	Coastal	Total	Coastal	Total	Coastal	Total	
Florida	7,659,364	9,746,320	10,501,222	13,527,968	37	39	78
U.S. Total	31,340,808	226,546,368	36,061,500	254,293,104	15	12	14

Source: National Planning Data Corporation, U.S. Census Bureau (IIPLR, 1995).

Florida accounts for the largest share of insured coastal exposures in the country. Table 5-4 shows that during the 13-year period from 1980 to 1993 Florida exposures increased from $332.9 billion to $871.7 billion. At that rate of growth and development, the value of property at risk in the state will shortly pass $1 trillion (IIPLR, 1995). See Figure 5-1 for a map of residential and commercial exposures in the state.

When looking at exposure values, it is of interest to note that some of the individual counties in Florida have significantly more property at risk than the total exposures for 16 of the 18 Atlantic and Gulf coastal states. For example, in 1993 Dade County (where Miami is located) exposures totaled $160.8 billion. Just to the north, Broward and Palm Beach counties had exposures of $116.4 billion and $103.0 billion, respectively. Thus, within a fairly small geographic area in southeast Florida, highly vulnerable to hurricanes, there is a concentration of insured property exposures approaching $400 billion (IIPLR, 1995).

Estimating Future Losses

Increased coastal exposures and the magnitude of damage caused by Hurricane Andrew have led to major upward revisions in projections of damage from future severe hurricanes. In addition, post–Hurricane Andrew investigations revealed that property losses were exacerbated by noncompliance with building codes, faulty structural designs, and the scarcity of building material and tradesmen to make repairs (Dade County Grand Jury, 1992). Consequently, most insurers, meteorologists, and academics, who had believed it unlikely that any one U.S. hurricane could cause insured damages on the scale of Hurricane Andrew, altered their views. According to A. M. Best, "Before 1992, many weather experts believed that the worst-case hurricane in the U.S. would produce less than $10 billion in insured property damage" (BestWeek Property/Casualty Supplement, 1996). That view has changed dramatically.

Recent estimates by Applied Insurance Research (AIR) suggest that if a Category 5 hurricane were to hit Miami or Ft. Lauderdale, it would cause more than $51 billion in insured losses. Factoring in additional costs such as uninsured direct and indirect losses, flood insurance losses, or other economic losses not insured by the private insurance industry would make the total cost much higher (IIPLR, 1995).

The insurance industry has reassessed damageability functions for structures subject to hurricane damage and has made adjustments to re-

TABLE 5-4 Value of Insured Coastal Property Exposures for Florida, 1980–1993 (in Millions of Dollars)

Location	Residential		Commercial		Total		Percent Change 1980–1993
	1980	1993	1980	1993	1980	1993	
Florida	177,709	418,392	155,213	453,288	332,922	871,680	162
U.S. Coastal Total	615,598	1,639,741	514,066	1,507,304	1,129,664	3,147,045	179
U.S. Total	4,240,948	10,278,875	3,807,860	11,043,124	8,048,808	21,421,999	166

Source: Applied Insurance Research, Inc. (IIPLR, 1995).

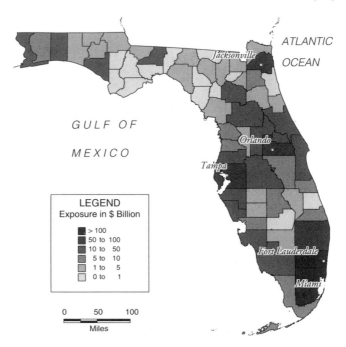

FIGURE 5-1   Total and residential commercial exposures in Florida.

flect building code noncompliance, fraud in construction, and the post-loss increase in the costs of labor and materials. The industry has also expanded use of, and reliance upon, unproven loss estimation models, because of their ability to handle complex damageability curves and other data. The credibility of these models will be tested when the next major hurricanes strike.

HURRICANE ANDREW AND THE
FLORIDA INSURANCE MARKET

On August 24, 1992, Andrew, a Category 4 hurricane, hit the southern coast of Florida just south of Miami, causing economic damages estimated in excess of $25 billion, which makes it the most expensive hurricane ever to strike the United States. The disaster caused an estimated $15.5 billion in insured losses according to Property Claim Services (PCS). (This figure does not include loss adjustment expenses.) Table 5-5 displays the insured damage by lines of coverage. According to the Florida Department of Insurance, "personal insurance policies covering residential and personal property loss were affected by the total

TABLE 5-5 Insured Hurricane Andrew Damage in Florida, by Line

Line of Insurance	Insured Losses[a]	Number of Direct Claims
Homeowners'	9,762,571	280,000
Commercial Multiperil	3,309,863	50,517
Fire and Allied Lines	932,952	24,467
Auto, Physical Damage	320,786	161,400
Mobile Home Owners'	179,472	11,779
Farm Owners'	13,905	1,245
Other Lines	491,530	17,177
Total	15,018,265	680,239

[a]In thousands of dollars.

Note: Individual dollar losses and claims by line do not add to totals because not all companies reported loss information by line of insurance.

Source: Florida Insurance Department. Based on information from insurers writing 97 percent of property-casualty insurance in the state, as of December 31, 1992.

destruction of over 28,000 homes; damage to 107,380 homes; and 180,000 persons made homeless" (Gallagher, 1993, p. 1). The large number of homeowners' claims was due to damage to contents and other covered items for owner and non-owner occupied structures. At least 11,779 mobile home claims were made, many involving total loss of the home and its contents (IILPR, 1995). Commercial insurance claims covering property damage, inventory loss, and loss of use were triggered by the total destruction of, or damage to, 82,000 businesses. The loss of power to 1.4 million residents and the loss of telephone service to some 80,000 locations resulted in the interruption of businesses, the loss of electronic records, and the spoilage of food (Gallagher, 1993).

The short-term impact of this storm on the socioeconomic structure of southern Dade County has proven significant. In the year following Andrew, at least 50,000 people decided to move out of the area permanently (IIPLR, 1995). The Metro-Dade Planning Department estimates the total restoration of the population and economic base of the affected area will not occur until the year 2000 (IIPLR, 1995).

Immediate Market Impacts

At the time of Hurricane Andrew, the Florida insurance market ranked fifth in size among the states, with 4.9 percent of total property-casualty premiums (King, 1993). Due to the magnitude of insured losses inflicted by Hurricane Andrew, an immediate post-storm reaction of a number of insurance companies was to attempt to reduce their underwriting exposure. In early 1993, the Florida Department of Insurance released a list showing 39 insurers who intended either to cancel or not renew 844,433 policies in Florida's post-Andrew marketplace. Factors that influenced the private insurers to take such actions included:

- the inability to obtain adequate reinsurance for new and existing risks in Florida: when available, the cost of reinsurance was considered too high by primary insurers relative to their alternative costs of capital, and the price they could obtain for coverage from the insureds;
- new information from catastrophe risk models indicating that existing levels of exposure might be more significant than previously realized, and that these exposure levels were disproportionate to individual company and industry financial resources when considered in the aggregate;
- significant reductions in insurers' policyholders surplus as a result of Hurricane Andrew losses;
- concerns about rate adequacy, especially for coastal counties, and for certain classifications of risk such as condominiums;
- "hidden" exposures resulting from potential assessments by various other mandated insurance mechanisms (e.g., residual markets, catastrophe funds, guarantee funds, etc.);
- fear that an unfavorable catastrophe exposure would negatively impact the rating by agencies such as A. M. Best and Standard and Poor's.

Some insurers have determined their potential risk of a catastrophic loss in Florida to be so great that they have withdrawn from the state entirely. Withdrawal was prompted in some cases by concerns on the part of the regulator in the insurer's home state about solvency, about the ability to meet other obligations outside Florida, and about the effect of steeply increased reinsurance costs.

Institutional Arrangements

In order to address the needs of stranded policyholders, the insurance commissioner issued an emergency rule on October 15, 1992, that activated the Florida Property and Casualty Joint Underwriting Association (FPCJUA) on a temporary basis to provide property coverage to all policyholders of insolvent insurers who were already making repairs or were planning to make repairs to their homes. A special session of the legislature held in December 1992 ratified the actions of the Department of Insurance (DOI) and extended the coverage until repairs were completed and the homes became eligible for insurance in the voluntary market.

The FPCJUA had been created in 1986 as a vehicle to provide commercial coverage if and when such coverage became unavailable in the voluntary market. Even though activation of the FPCJUA for the purpose of issuing residential coverage clearly had a questionable legal basis, the DOI initiated this action to resolve a short-term problem and secure the support of private insurers, who agreed not to challenge the emergency rule after it was issued.

The Florida Insurance Guaranty Fund (FIGA) had been created in 1970 to cover the claims payments of insolvent insurers. When insolvencies occurred, FIGA was funded by assessing the property premiums written by all other insurers in the state based on their percentage of market share. The maximum assessment allowed by law at the time of Andrew's occurrence was 2 percent of gross annual premiums, generating about $70 million in available funds (State of Florida, 1993).

Since FIGA's creation, assessments had been minimal and infrequent. However, the liabilities of the Hurricane Andrew insolvencies—more than $400 million in unpaid claims—required a doubling of the assessment rate on insurers (Conning and Company, 1994). Even at this assessment level, FIGA was still short of the funds needed to respond to the insolvencies. In order to fill this gap, the Florida legislature in the December 1992 special session authorized FIGA to issue up to $500 million in tax-exempt revenue bonds. These were issued in February 1993 through the city of Homestead as tax-free municipal bonds, and about $430 million in claims were paid through June 1993.

To repay these bonds, the state legislature authorized the insurers to collect a special additional assessment on policies of 2 percent of premium per year for the next ten years (IIPLR, 1995). The bonds were also insured by MBIA, Inc., the municipal bond insurer. By 1997, with sufficient funds at hand, FIGA retired the city of Homestead's bonds ahead of schedule.

Despite Florida's actions, which successfully dealt with the post-Andrew insolvency crisis, it was obvious to most that problems associated with insolvencies, including delays in claims payments from FIGA and the burden of continuing indebtedness, created an unstable situation which could grow worse if another major catastrophe struck Florida in the near term. Both the state legislature and the DOI realized that Florida needed a more permanent insurance mechanism to both prevent massive insolvencies and to more efficiently handle those that occur.

Insolvencies

Nine property-casualty insurance companies became insolvent as a direct result of Hurricane Andrew (King, 1993). Eight of these companies were domiciled in the state of Florida, with individual surplus amounts ranging between $1.4 million and $4.4 million. The ninth company, with a surplus of $24 million, was domiciled in Oklahoma but wrote most of its policies in Florida.[3] While there was no single characteristic common to these companies, in general they had all experienced poor underwriting results in the year or two prior to Andrew, and most had realized a depletion in their surplus ranging between 9 percent to 68 percent. Six of the insurers were relatively new to the industry, having been incorporated in the mid-1980s (A. M. Best, 1992).

These insolvencies added to the financial burden of other insurers when they were assessed by the FIGA to pay the losses of the defunct companies up to the maximum provided in the law. In fact, a tenth company became insolvent as a result of FIGA's post-Andrew assessments. [This was American Property and Casualty Company, with $2.1 million in surplus (Conning and Company, 1994).]

Insolvencies also contributed to problems with the availability of property insurance after Andrew. Florida law, as it existed at that time, required that every policy issued by an insolvent insurer be canceled within 30 days to allow the policyholders time to find replacement coverage. Following the Andrew-related insolvencies, however, policyholders with damaged property were unable to purchase a policy because most insurers do not accept damaged property under a new policy (Gallagher, 1993).

---

[3]The companies domiciled in Florida were: First Southern ($2.363 million in surplus); Florida Fire & Casualty ($1.386 million); Great Republic ($2.524 million); Guardian Property & Casualty ($2.192 million); Insurance Company of Florida ($4.353 million); Nova Southern ($1.966 million); Ocean Casualty ($2.547 million); and Regency Insurance Company ($1.418 million). The Oklahoma company was MCA Insurance Company with a surplus of $24 million (King, 1993, p. 4; Conning and Company, pp. 57 and 147).

SEEKING SOLUTIONS

Amid the public clamor for claims payments from FIGA and the unavailability of insurance from the private sector, the Florida Department of Insurance, the state legislature, and the insurance companies (which had previously competed fiercely for business) joined forces to address the mounting public demands for affordable insurance available to all qualified homeowners and commercial establishments. In the ensuing months and years, the regulatory and legislative process generated a latticework of interlocking measures to support the market and make affordable coverage available once more.

Regulatory Response

The State of Florida operates under a "plural executive" concept in which the executive power is divided between the governor and six other elected magistrates. One of the elected magistrates is the treasurer and insurance commissioner, who under Florida statutes has both the power to regulate insurance in the state and also the exclusive executive authority to oversee the insurance industry.

Chapter 627 of the Florida Statutes specifically gives the Department of Insurance (DOI) the authority and responsibility to regulate insurance rates. Insurance companies must file rate requests with the DOI, which are then subject to review and approval or modification. This review uses three basic criteria: rates must not be excessive, inadequate, or unfairly discriminatory.

In order to evaluate whether rates meet these criteria, the DOI uses various factors including projected future losses, the degree of market competition, reinsurance costs, and the catastrophe loss provision (State of Florida, 1993). After Hurricane Andrew, the question of rate adequacy, whether in voluntary or residual markets, has been paramount in the interaction between the insurance commissioner and the state's property insurers.

Moratorium on Insurer Withdrawals

As companies indicated their intent to either reduce their writings or leave the Florida market entirely, the Florida insurance commissioner issued an emergency rule (4ER92-11) on November 17, 1992, which delayed insurance companies from withdrawing from Florida for 90 days.

Mobile home park flattened by Hurricane Andrew, 1992 (FEMA).

They were required to file a written statement of intent containing details of the reasons for their actions 90 days before taking action and had to be prepared to demonstrate that their withdrawal would not adversely affect the market (Florida Department of Insurance, 1994). When the emergency rule was set to expire on February 15, 1993, the DOI issued another emergency rule (4ER93-5) extending the regulations pertaining to withdrawal for an additional 90 days until May 12.

After the second emergency rule expired, the DOI took a more drastic step; on May 19, 1993, the DOI imposed a moratorium on the cancellation and nonrenewal of residential property policies for 90 days to prevent insurers from carrying out their proposed cancellations at the start of the new hurricane season. Shortly thereafter, during the special session of the Florida legislature in the last week of May, the legislature enacted a law extending the moratorium until November 14, 1993, so that a legislatively created Study Commission on Property Insurance and Reinsurance could look into the viability of the property insurance industry and the adequacy of reinsurance.

During a third special session of the legislature, held in November 1993, the state legislature enacted a bill requiring a phaseout of the moratorium, which would last three years, from November 14, 1993 to November 14, 1996. The law provided that in this phaseout period individual insurers could not cancel more than 10 percent of their homeowners' policies in any county in one year, and that they could not cancel more than 5 percent of their property owners' policies statewide for each year the moratorium was in effect (Insurance Services Office, 1994). The moratorium was, at best, a stopgap measure, designed to buy time while long-term solutions to Florida's property insurance problems were developed (State of Florida, 1993).

During the 1996 state legislative session, the moratorium phaseout law was extended for an additional three years until June 1, 1999. The moratorium's application was broadened to include condominium associations' master policies in addition to homeowner policies (Briggs, 1996). However, the legislature also allowed accelerated nonrenewals. An insurer, with approval of the commissioner, could use its entire "quota" of nonrenewals allowed through June 1, 1999, in the first year. Several major insurers filed for permission to do that, and those filings were approved.

Recently, the U.S. District Court for the Northern District of Florida ruled that the implementation of the moratorium by the state legislature in 1993 was within the state's constitutional rights, considering the economic crisis that could have been caused if a large number of insurers had left the state after Hurricane Andrew in 1992. The ruling is being appealed by the plaintiff insurer. The 11th Circuit of Appeals heard an oral argument in February 1998, but as of this writing has not ruled on it (Jack E. Nicholson, Chief Operating Officer, Florida Hurricane Catastrophe Fund, personal communication, 1998).

While the moratorium serves as a barrier to market exit, it is also a barrier to market entry. Insurers not doing business in Florida are reluctant to enter because they are concerned that they may not be able to exit if business circumstances change. More significantly, the moratorium contributes to the reluctance of insurers already writing in the state to increase their market share. Also contributing to this reluctance are the formulae of the residual market mechanisms, which base assessments on a carrier's market share.

A State Hurricane Trust Fund

In the special legislative session held in November 1993, the Florida Hurricane Catastrophe Fund (FHCF) was created to relieve pressure on insurers to reduce their exposures to catastrophic hurricane losses. This fund enables insurers to remain solvent while renewing most of their policies scheduled for nonrenewal (Florida House of Representatives, 1993); it reimburses a portion of insurers' losses following major hurricanes. The FHCF, a tax-exempt trust fund administered by the State of Florida, is funded by premiums paid by insurers that write insurance policies on both personal residential property and commercial residential property (defined as apartments and condominiums). Participating insurers are charged premiums based on their property exposure in the state. Rates are based on type of coverage (personal residential, commercial residential, or mobile home), zip code location grouping, construction type, and deductible level.

The FHCF is never obligated to pay more than its assets and borrowing capacity. At year-end 1997, the fund had an estimated capacity of $8 billion—approximately $2 billion in cash derived from premiums and interest earnings, and a borrowing capacity estimated at $6 billion (Jack E. Nicholson, FHCF, 1998). It was planned that these funds would be raised through the issuance of revenue bonds funded by emergency assessments of up to 4 percent of premiums on all property and casualty insurers, excluding workers' compensation writers. As of the spring of 1998, the State of Florida was seeking a private letter ruling from the Internal Revenue Service regarding the ability of the FHCF Finance Corporation to issue tax-exempt debt. Should the ruling be obtained, the estimated bonding capacity of the FHCF will increase around $2 billion due to the lower cost of financing on a tax-exempt basis.

If the fund does not have sufficient resources available to pay all claims in full, each insurer will first be eligible for reimbursement up to an amount equal to the insurer's share of the actual premium paid for that contract year multiplied by the actual claims paying capacity. Each insurer is therefore entitled to its share of the fund's capacity based on the premiums it has paid to the fund. Stated another way, a coverage multiple is calculated by dividing the total claims-paying capacity by FHCF aggregate premiums for the particular year. The product of the coverage multiple and each insurer's FHCF reimbursement premium will determine each insurer's expected minimum payout from the fund. Once insurers are reimbursed for their losses up to their coverage level, reimbursements are next based on losses subject to available funds. Losses are then reimbursed on a prorated basis at the highest level possible, given the remaining available funds.

The coverage multiple for 1997 was calculated by dividing $7.97 billion, which is the sum of the projected fund balance at year-end plus the estimated bonding capacity ($1.97 billion + $6 billion), by the 1977 FHCF reimbursement premiums of $471 million, yielding a multiple of 16.9. This coverage multiple times an insurer's 1997 FHCF reimbursement premium produces an expected minimum payout from the fund. For example, an insurer that paid a $1 million premium could expect to recover $16.9 million from the FHCF. The insurer may recover more depending on whether other insurers utilized their share of the FHCF's capacity (Jack E. Nicholson, FHCF, personal communication, 1998).

The tax-exempt status of the FHCF is a significant advantage to the insurers participating in it, as it removes a level of potential income taxation resulting from the annual buildup of contingent reserves in years when there are few, or no hurricanes, and provides for a dedicated accumulation of funds in Florida for the benefit of the state's policyholders. Some insurers have now taken steps to issue bonds in the capital markets prior to future hurricane events. In June 1997, USAA successfully floated a catastrophe bond. This bond has two layers of debt: one is subject to interest forgiveness should USAA suffer a loss in excess of $1 billion from a Category 3, 4, or 5 hurricane; the second layer has both principal and interest at risk. The targeted capacity of $400 million was oversubscribed partly because investors are now more familiar with these types of instruments, but also because of the very high return on the investment (Doherty, 1997).

RESIDUAL MARKET MECHANISMS

Residual market mechanisms (RMMs) provide "insurance of last resort" to individuals who are unable to obtain coverage in the voluntary market. As the example of the Florida Hurricane Catastrophe Fund illustrates, these mechanisms do not actually "remove" losses from the private insurers' financial burden. All insurers writing in the state of Florida must participate in these mechanisms and are subject to assessments by them. Thus, less desirable risks are borne in proportion to a company's market share in the relevant lines of insurance.

Two property insurance residual market mechanisms now exist in Florida, the Residential Property and Casualty Joint Underwriting Association (RPCJUA) and the Florida Windstorm Underwriting Association (FWUA).

Residential Property and Casualty Joint Underwriting Association (RPCJUA)

The RPCJUA was created in the December 1992 special legislative session to address the unavailability of property insurance in the voluntary market for homeowners' and dwelling fire coverage. [It included the Florida Property and Casualty Joint Underwriting Association (FPCJUA) referred to earlier in this chapter.] It was organized in January 1993, and began issuing policies in March, 1993. The policies issued by the RPCJUA cover single-family homes, townhomes, mobile homes, individual condominium units, and the contents of rented apartments and homes. These policies may be written as complete multiperil policies, which include wind coverage, or are written to exclude wind coverage in the areas eligible for coverage through the Florida Windstorm Underwriting Association.

The section of state law addressing rates has been amended over time. In 1995, lawmakers required the RPCJUA to base its homeowners' rates on the rates and statewide market shares of the 10 largest private insurance companies in the state. In each county, the average rate was required to be as high or higher than the highest average county rate charged by one of those 10 companies. For mobile homes, the rate is set as high or higher in each county than the highest average county rate

---

SIDEBAR 5-1 ▬▬

Current Status of the Residential Property and Casualty Joint Underwriting Association (RPCJUA)

As of February 28, 1998, the RPCJUA stood at 372,289 personal lines residential policies, with $50.6 billion in coverage. Much of the exposure in the personal lines residential book is in south Florida. Approximately 271,000 policies, representing 73 percent of the total, and $42 billion in exposure, representing 83 percent of the total, were in three south Florida counties: Dade, Broward, and Palm Beach. On the commercial lines residential side, there has also been a significant drop in policy count and exposure. The commercial book peaked at 2,291 policies in March 1997, with $11.5 billion in coverage. As of March 21, 1998, the commercial book stood at 1,041 policies and $4.7 billion in coverage.

There has been continuing concern among insurers and legislators about the RPCJUA's fiscal condition and the aggregate windstorm exposure. For this reason the state legislature created special purpose homeowners' insurance compa-

charged by one of the five largest private companies writing mobile home policies, based on the companies' statewide market shares. The higher price of RPCJUA policies, therefore, encourages consumers to seek private market coverage (Gallagher, 1993).

In the early history of the RPCJUA, lawmakers and state regulators viewed its rates as too low and undercutting the rates charged by private insurance companies. This fact, at least in part, helped spur the association's early growth. However, between 1993 and 1998, the total cumulative increase in premium on a statewide average basis was 90 percent for homeowners' policies, 83 percent for dwelling fire policies, and 67 percent for mobile home policies. Today, the association's rates are generally the highest in the market in each of Florida's 67 counties.

The RPCJUA accepts almost all applications for coverage, although it is not allowed to write windstorm coverage in areas eligible for coverage by the Florida Windstorm Underwriting Association. The RPCJUA peaked at 937,000 personal residential lines policies in September 1996, with $98.2 billion in coverage. However, the association has been sig-

---

nies for RPCJUA takeouts. In late 1997, the first special purpose homeowners insurance company was formed under Florida law. Companies such as this one may only write policies removed from the RPCJUA or assumed from an unaffiliated authorized insurer and cannot write policies in the voluntary market. They are not members of the RPCJUA or the FWUA and therefore not subject to regular assessments. However, their policyholders are still subject to emergency assessments. These insurers would be subject to assessment only by the guarantee fund, the FIGA, and the catastrophe fund.

In May, 1997, the RPCJUA board put into place a catastrophe-financing program that essentially doubled the association's capacity to pay claims from a major hurricane. The program includes $279 million in regular assessments upon member companies, $1.15 billion in proceeds from the Florida Hurricane Catastrophe Fund in 1997, $500 million in pre-event notes (which are invested in a trust fund until needed), and a $1.5 million line of credit with a worldwide group of banks. (This financing program replaced the $1.5 billion line of credit the RPCJUA put in place in 1995.) This new financing program, coupled with the

continued

nificantly reduced in size since then through various depopulation initiatives.

Under the law, most property insurance companies (except for special purpose homeowners' companies) are required to be members of the RPCJUA. Each year, the association determines the companies' market shares by calculating the amount of premium they wrote in the preceding calendar year. If the association incurs a deficit, the RPCJUA can impose "regular assessments" upon each of the member companies, based on their market shares. The companies must pay this assessment to the RPCJUA to cover the association's expenses. These regular assessments are capped at either 10 percent of the deficit or 10 percent of the total amount of statewide written premiums that private insurance companies wrote for the previous year, whichever is higher.

The insurance companies can recoup these regular assessments from their policyholders by filing for a rate increase with the Florida Department of Insurance. If a deficit exceeds the amount that can be covered through a regular assessment, the RPCJUA can impose "emergency assessments" on all property insurance policyholders in Florida. The poli-

SIDEBAR 5-1 *Continued* ━━━━━━━━━━━━━━━━━━━━━━━━━

significant reduction in the number of policies in the RPCJUA, has enabled the association to have sufficient financing in place to cover a one-in-100-year hurricane.

In the fall of 1996, when the RPCJUA stood at nearly 937,000 policies, the association's probable maximum loss from a one-in-100-year hurricane stood at $4.8 billion. As of February 1998, when the RPCJUA had been reduced to 372,000 policies, the one-in-100-year probable maximum loss was down to $2.75 billion.

The law establishing the RCPJUA sets out some specific incentives for companies to remove policies from the association. The statute also gives the RPCJUA board of governors more general authority to establish additional incentives to hasten depopulation. The board has used this authority in crafting various depopulation programs, which have had a considerable impact on reducing the size of the RPCJUA. Had these depopulation efforts not been undertaken, the association estimates that it would have grown to 1.5 million or more policies, which would have made the RPCJUA the largest property insurer in the state. As of the end of February 1998, 22 companies had in force 676,797 policies that they had

cyholders would be required to pay the assessments to their insurance companies, who would then turn the money over to the RPCJUA.

The emergency assessments are capped annually at 10 percent of the deficit, or 10 percent of the total amount of property insurance premium that private insurers wrote in Florida in the previous year, whichever is larger, plus interest, fees, commissions, required reserves, and other costs associated with financing the original deficit. The RPCJUA has imposed regular assessments for deficits totaling $17.7 million in 1994 and $22.8 million in 1995 (Adams, 1996). The RPCJUA recorded a modest surplus in 1996. In 1997, the surplus was $170.2 million on a GAAP reporting basis (GAAP = generally accepted accounting principles). The 1997 surplus was primarily the result of a one-time takedown of loss reserves in 1997 and the fact that no hurricanes hit Florida in 1996 and 1997. To date, the association has not imposed an emergency assessment.

Rate adequacy and the size of the residual market mechanisms have been an ongoing concern to the legislature. When the legislature could not agree on revisions to these mechanisms, it created the Legislative

---

removed from the association. The number of policies removed by these companies is actually higher, but in some instances the policies did not ultimately wind up with the companies—that is, policyholders shopped for coverage on their own, left the state, or chose to go without coverage, and so on.

While the depopulation efforts have been successful, the RPCJUA continues to receive a heavy flow of new business each month that threatens to erode the success the association has achieved in reducing its policy count. As of early 1998, the RPCJUA was writing approximately 15,000 to 16,000 new policies monthly. As a result, the association has embarked on "keep out" programs designed to prevent policies from ever entering the RPCJUA. As a first step, the RPCJUA has solicited expressions of interest from insurers willing to write a meaningful amount of new business each month. The companies have been asked to provide information on the types of policies they are willing to write. Association producers have been requested to voluntarily submit applications for coverage to the companies for their underwriting review. To date, 11 companies have expressed interest in participating in these programs.

Working Group on Residual Property Insurance Markets to fully examine the residual markets and make recommendations for legislative changes in the 1997 session. The primary purpose of the Legislative Working Group on Residual Property Insurance Markets was to develop recommendations for the legislature for a permanent residual market mechanism to replace the existing RPCJUA and the FWUA. The Working Group issued a final report in December 1996, recommending that the RPCJUA and the FWUA should remain separate, but work toward better coordination and eventual consolidation. The report said the boards of the two associations should develop a transitional plan for merging the two entities into a single residual market when the RPCJUA reaches 100,000 policies. The Legislative Working Group did not address the rate issues.

In an effort to allay concerns about its liquidity and its ability to pay claims after another hurricane, the association's board of governors in 1995 secured a $1.5 billion line of credit from a consortium of banks. In 1995, the board also authorized the purchase of catastrophe reinsurance at a time when several hurricanes simultaneously formed in the Atlantic and appeared to pose a threat to Florida. At that time, the board approved the purchase of $300 million in second-event coverage attaching after an aggregate of $1.175 billion in losses. In September 1997, the association purchased facultative reinsurance, excluding wind, for its commercial book of business covering locations valued in excess of $10 million. (See Chapter 2 for a discussion of different reinsurance treaties, including facultative.)

Because of an unintended problem created by wording in the 1996 special purpose homeowners' insurance company legislation, the RPCJUA would have been in default under its 1995 $1.5 billion line of credit if the Florida Department of Insurance (DOI) had allowed a special purpose homeowners' insurance company to form. Therefore the DOI issued an order stating that no such companies would be allowed as long as the possibility of a default existed. During the 1997 session, the Florida legislature amended the special purpose homeowners' insurance company statute and corrected the wording problem. As a result, the DOI order was rescinded.

Florida Windstorm Underwriting Association (FWUA)

This residual market mechanism, in existence since 1970, is the oldest in the state. It was created to assure the availability of windstorm and hail insurance coverage for dwellings and other structures in designated eligible areas. The eligible areas are determined by the DOI and have traditionally been areas at high risk for wind damage (Gallagher, 1993).

If an insurer operates at a loss, the FWUA, like other residual market mechanisms, assesses those insurers operating in Florida. Assessments are based on a company's market share for the previous year, but can be reduced by credits derived from voluntarily writing wind coverage in FWUA-eligible areas. The total assessment for member companies may not exceed 10 percent of its statewide property premium writings in a given year.

The FWUA's loss exposure and the number of its policies in force have grown dramatically since Hurricane Andrew. From December 1992 to December 1994, policies in force increased by 198 percent. As of the end of 1996, loss exposure was estimated at $49 billion on 320,000 policies in force (PIPSO Reports, 1997). The growth in exposure is directly linked to the additional areas in Florida in which FWUA has begun to operate. In the aftermath of Hurricane Andrew, the DOI designated Dade and Broward Counties eligible for FWUA coverage. In the subse-

Store destroyed by Hurricane Andrew, 1992 (FEMA).

quent months, the department also made other coastal areas eligible, which greatly increased the FWUA's exposure.

During the 1995 hurricane season the FWUA received about 9,300 claims caused by Hurricane Opal. These amounted to $145 million in losses, with approximately $70 million of that total reimbursed by reinsurance. The state's insurers were assessed $84 million for Hurricane Opal, which followed an October 1995 assessment of $33 million to cover Hurricane Erin losses (information provided via facsimile, FWUA, 1998).

In the wake of these economic impacts, the FWUA has reevaluated and determined that its fully indicated rate level should be 124.3 percent of the existing base rate. However, recognizing that policyholders need time for rate increases, the FWUA requested that the indicated rate be phased in gradually. As a result, the FWUA requested an overall rate increase of 67 percent. In March 1996, the Department of Insurance approved only a 31 percent rate increase effective with new and renewal policies on or after June 1, 1996. In August 1997 the FWUA filed a rate increase averaging 62 percent statewide, to be phased in over a period of three years (Adams, 1997). The insurance commissioner rejected the hike, however an independent arbitration panel recently overruled him and indicated that the FWUA can raise rates an average of 12 percent after August 1, 1998. Dade and Broward County rates could go up as much as 40 percent, which is the maximum increase allowed under the ruling (DeLollis, 1998).

THE REINSURANCE MARKET[4]

In the four years prior to Hurricane Andrew the world reinsurance market sustained a number of catastrophic losses in Europe, in the North Sea, and in Asia. Thus, the profound effect of Hurricane Andrew was felt in a market segment already suffering losses. The major impacts for primary company buyers of reinsurance in the year following Andrew included a severe shortage of catastrophe property reinsurance capacity and stricter policy terms and conditions, as well as sharp increases in property catastrophe cover rates due to a significantly increased demand against scarce supply.

These profound changes in the reinsurance market further reduced

---

[4]The help and assistance of Guy Carpenter & Company, Inc., New York, N.Y., in providing access to background information for this section is gratefully acknowledged.

A Florida hurricane (FEMA).

the primary carriers' willingness to write additional business, or even renew portfolios in force. The primary carriers were faced with an inability to pass along the increased reinsurance costs, and a concurrent inability to reduce or effectively manage catastrophe exposures. Under these conditions, the reinsurance market was viewed by the public and the Florida Department of Insurance as part of the problem rather than as a hoped-for solution. Reinsurers for their part were paying losses far beyond their expectations. It is doubtful that, as the magnitude of the Andrew loss became clear, the reinsurance market could have contributed significantly to a marketplace solution because of its own inability to diversify away the concentrations of Florida exposures in portfolios, even on a worldwide basis.

In response to the acute distress in the catastrophe reinsurance market, reinsurance intermediaries, aided by investment bankers, raised capital to create new reinsurers in Bermuda, virtually overnight. As pointed out in Chapter 2, more than $4.8 billion in private capital was raised, and in the course of 1994 eight monoline catastrophe reinsurers were formed (wind and earthquake coverage only). U.S. insurers and reinsurers were major backers of these new companies. The market opportunity available to new entrants was huge, and Bermuda offered an excellent regulatory and tax-free climate in which to conduct the highly volatile catastrophe business.

While concerns were voiced initially about the durability of these new monoline catastrophe companies, the companies have proven to be successful, accounting for approximately 32 percent of worldwide catastrophe capacity in 1995; the U.S. markets provided 25 percent (Piper, 1995). It is worth noting that several of the Bermuda reinsurers have begun offering additional lines of coverage (liability excess, for example), either directly or through mergers or acquisitions.

In conjunction with limited post-Andrew capacity, reassessment of exposures and risk management programs by primary insurers and reinsurers have also stimulated exploration of other means of financing catastrophic coverage. In 1994 and 1995, reinsurers, investment bankers, and financial market traders moved to develop contingent capital, reinsurance, and derivative risk management products to add risk-bearing capacity from the capital markets. Depending upon the buyers' needs and willingness to accept basic risk, these new risk management products will provide an alternative, or a supplement, to traditional reinsurance (Reinsurance Association of America, no date). A key mechanism for the introduction of the capital markets is expected to be the creation of an index of catastrophe loss which will provide transparency to the capital markets investor. Much work and education remains to be done with investors, who must accept insurance risk as an asset class for inclusion in portfolios.

THE FUTURE

Each hurricane season highlights Florida's extreme vulnerability to the windstorm peril. Issues of insurance availability and affordability, as well as the solvency of primary insurers, remain the major concern of insurers, the state's insurance regulator, its legislature, and its insurance consumers. All the players are not necessarily in agreement as to how these elements are best balanced, however. It is clear that the basic problems brought to light by Hurricane Andrew remain the primary challenge six years later.

While this chapter concentrates on insurance mechanisms in the post-Andrew environment, it is important to note that they are not the only factors in improving Florida's lot. New ways to finance catastrophe losses (whether private or government), new risk management tools, and new insurance products must be met on equal terms by thorough disaster planning, mitigation programs, public education, and a stronger built environment. Initiatives are currently under way that recognize the need for an approach where each stakeholder plays its role.

One such initiative is a proposal by the Florida Insurance Council (FIC) for an agreement establishing post-hurricane cooperation between the FIC, the state insurance department, and state and local emergency management officials. In addition, the governor created a Building Codes Study Commission to evaluate the current effectiveness of the Florida building code system and recommend any necessary reforms (Executive Order 96-234). In early 1998 this commission recommended that Florida should have one building code for statewide use which would be administered and enforced by local government building and fire officials (Five Foundations for a Better Built Environment, 1997).

Another initiative is the Florida Commission on Hurricane Loss Projection Methodology which was established by the state legislature in 1995 to examine the role of computer models in determining insurance rates. The commission developed a set of standards which a model must meet and hired a professional team to audit the modeling companies on site. By December 1997 the three principal models (those of Applied Insurance Research, EQECAT, and Risk Management Solutions) had all met the commission's standards and are approved for use in establishing insurance rates. The extent to which the Department of Insurance approves rate filings based on computer models is still unsettled (Mark Johnson, Director of the Institute of Statistics, University of Central Florida, personal communication, 1997).

The extent to which all of Florida's stakeholders come together and create successful collaborations will, in large measure, determine the state's ability to deal with the next Hurricane Andrew when it arrives, as it inevitably will. In the interim there will be ongoing tensions between the insuring public, the regulator, the legislature, and the insurance industry on the benefits and economic value of the changes already implemented, or under consideration. The economic future of the state of Florida, and its competitiveness, is at stake.

CHAPTER SIX

The National Flood Insurance Program

EDWARD T. PASTERICK

F LOODS, LIKE OTHER NATURAL disasters, present us with two challenges: how to contain the cost of the damage they cause, and how to provide economically feasible relief to victims that will help them recover from the disaster. Containing the cost of damage caused by flooding has generally been accomplished either through constructing flood control works designed to keep floodwaters away from properties located in the floodplain, or through land use regulation employed to guide construction away from the path of floods or ensure safer building in the floodplain.

The federal government's first response in this arena was the Flood Control Act of 1936 (49 *Stat.*, 1570), which launched a national program of structural flood control works. Structural flood control measures have had their critics almost from the start. They argue that, while structural projects have afforded a degree of protection from floods, they also give the residents of floodplains a false sense of security, which has resulted in even further encroachment on the floodplain, thus increasing rather than reducing the cost of potential losses (White, 1953).

For assisting flood victims in the post-disaster recovery process, the only available recourse until 1968 was federal disaster assistance, which took the form of disaster loans and grants. However, the continuing increase in the costs to the federal treasury of providing this relief has caused policymakers to look closely at the feasibility of providing insurance coverage against flood losses as a preferable alternative to federal assistance.

When insurance against flood losses was first discussed in the 1950s, it became clear that private insurance companies could not profitably provide such coverage at an affordable price. This was primarily because of the catastrophic nature of flooding and the insurers' inability to develop an actuarial rate structure that could adequately reflect the risk to which flood-prone properties were exposed. The conclusion was that such an insurance program would require a substantial involvement by the federal government.

With passage of the Federal Flood Insurance Act of 1956, Congress proposed an experimental program designed to demonstrate that private sector provision of flood insurance was commercially feasible (42 *U.S. Code*, sec. 2401 et seq.). This experimental program was never implemented, largely due to a lack of support from the insurance industry and the above-mentioned difficulty in establishing a sound actuarial foundation for the program. (It should be noted that private insurers do write limited amounts of flood coverage, usually for commercial insureds, under Inland Marine and "Difference of Conditions" policies. The existence of the National Flood Insurance Program has now created an environment favorable to the provision of some private residential flood coverage.)

The next serious look into the feasibility of creating a program of insurance against flood loss took place in the aftermath of Hurricane Betsy, which struck the Gulf Coast in the summer of 1965, and the heavy flooding on the upper Mississippi River in that same year. The Southeast Hurricane Disaster Relief Act of 1965 (P.L. 89-339), which provided federal financial relief to the victims of Hurricane Betsy, authorized a study to explore alternative ways of providing aid to flood victims.

This study resulted in a 1966 report from the Secretary of Housing and Urban Development (HUD) entitled "Insurance and Other Programs for Financial Assistance to Flood Victims" (U.S. Department of Housing and Urban Development). This report proposed that a flood insurance program to replace reliance on public assistance to repair flood-damaged property would be feasible if it contained the following essential elements:

- accurate estimates of risk
- compensation to the risk bearer
- the possibility of some level of premium subsidy, if publicly desirable
- incentives to policyholders to reduce risks
- incentives to states and local governments for wise management of flood-prone areas
- continuous reappraisal.

It was on the basis of this report that Congress passed the National Flood Insurance Act of 1968 (see Figure 6-1), which created the National Flood Insurance Program (NFIP). Later legislative revisions have generally been based on one or more of the key elements just listed. These revisions—the Flood Disaster Protection Act of 1973 and the National Flood Insurance Reform Act of 1994—both addressed perceived program inadequacies arising from experience with the program up to that point in time. The measures adopted had generally been anticipated and discussed in the 1966 HUD report (see Figure 6-1 on NFIP legislation).

NATIONAL FLOOD INSURANCE ACT OF 1968

(P.L. 90-448, Title XIII, 42 U.S. Code, sec. 4001 et seq.)

- Created the National Flood Insurance Program
- Made flood insurance available in communities that agree to adopt and enforce floodplain management ordinances

FLOOD DISASTER PROTECTION ACT OF 1973

(P.L. 93-234, 42 *U.S. Code*, sec. 4001 et seq.)

- Made community participation in the NFIP a condition of eligibility for certain types of federal assistance
- Made the purchase of flood insurance a condition for federal and federally related mortgage loans in high-risk flood area.

NATIONAL FLOOD INSURANCE REFORM ACT OF 1994

(P.L. 103-325, 108 *Stat.*, 2255)

- Strengthened the mandatory purchase requirements of the 1973 act
- Created the Flood Mitigation Assistance Grant Program
- Revised the Standard Flood Insurance Policy to include Increased Cost of Compliance coverage
- Included the Community Rating System in the statute.

FIGURE 6-1   NFIP legislation.

The NFIP was originally placed under the authority of the Secretary of HUD, who delegated program authority to the administrator of the Federal Insurance Administration (FIA). In 1979, FIA and its programs were transferred to the newly created Federal Emergency Management Agency (FEMA).

STRUCTURE OF THE NATIONAL
FLOOD INSURANCE PROGRAM

The NFIP structure includes three essential components: risk identification, hazard mitigation (i.e., actions taken to protect people and property from the flood peril), and insurance. Effective integration of these three components requires cooperation between the federal government, state and local governments, and the private property insurance industry. The authorizing legislation defined a role for each of these, which in some cases has been altered over the program's history.

Risk Identification

One of the major obstacles preventing the private insurance industry from providing flood insurance was its inability to adequately identify all the areas throughout the country that were vulnerable to flood hazards and then to effectively define the nature and extent of the risk. Because of the costs associated with conducting the hydrological studies needed to secure this information and the nationwide scope of the effort, the legislation assigned this task of risk identification to the federal government. Under the program as originally designed, community eligibility was conditioned on the completion of a flood insurance rate study and on a community's adoption of the resultant Flood Insurance Rate Map (FIRM). The FIRM serves both as the guide to the community for purposes of floodplain regulation and overall land use decision making, and as the source of risk information for property insurance agents to accurately rate policies.

The length of time required to produce a FIRM, and the resultant delay in insurance availability, prompted Congress to authorize the Emergency Program in 1969. This measure enabled property owners in a participating community to purchase limited amounts of flood insurance at estimated rates until completion of the FIRM, at which point the community was converted to full coverage under the Regular Program, which uses actuarial rates. Until 1974, both community participation and individual purchase of insurance was completely voluntary, and the studies conducted and maps produced were only for those communities that had applied for participation in the program.

When Tropical Storm Agnes struck the Eastern seaboard in 1972, it became clear that many communities were either unaware of the serious flood risk to which they were exposed or had been unwilling to take the necessary measures to protect residents of the floodplain. Very few of the communities affected by the storm had applied for participation in the NFIP, and even in these participating communities owners of most flood-prone property opted not to purchase flood insurance. This lack of response to the NFIP perpetuated the reliance of flood victims on federal disaster assistance to finance their recovery.

Because of this minimal participation, the 1973 Flood Disaster Protection Act made the NFIP responsible for identifying all communities nationwide that contained areas at risk for serious flood hazard. The NFIP was also required to notify these communities of their choice of applying for participation in the NFIP or forgoing the availability of certain types of federal assistance in their community's floodplains. This NFIP effort identified over 21,000 flood-prone communities in the 50 states and the District of Columbia. It also identified one community each in American Samoa, Guam, Puerto Rico, Territory of the Pacific, and the Virgin Islands. As of March 1998, 18,760 communities, including one each in the territories, had joined the NFIP. Most flood-prone communities that have elected not to participate are communities whose areas of serious flood risk are either very small or have few if any structures. The dramatic increase in community participation stemming from this nationwide notification has had the further result of increasing the number of detailed flood studies that the government has been required to complete. Although the hazard identification process identified all communities having any areas subject to serious flood risk, regardless of the size of such areas, not all communities have required a detailed study in order to participate in the NFIP.

The NFIP conducts two general types of flood studies, approximate and detailed. Detailed studies are conducted for communities with developing areas—that is, areas where industrial, commercial, or residential growth is beginning and/or where subdivision is under way, and where the development or subdivision is likely to continue. Communities with minimal flood risk have been brought into the program without such a detailed study being conducted. Approximately 6,300 of the 18,760 participating communities are in this low-risk category. Through fiscal year 1997, the cost of this massive study effort has been about $1.154 billion.

For purposes of risk identification, the NFIP uses as the standard of risk the "100-year flood," also referred to as the one percent flood or the base flood. This is a degree of flooding that has a one percent chance of occurring in a given year. On most community FIRMs the 100-year flood is indicated in terms of flood elevations in feet relative to mean sea level. In coastal areas, which are exposed to the additional hazard of wave action, the base flood includes the height of the waves above the stillwater elevation. The term "100-year flood" is problematic for the NFIP. It is a term of convenience intended to convey probability but has had the adverse effect of giving floodplain residents, who tend to interpret it in chronological terms, a false sense of security.

Hazard Mitigation

No program of insurance against flood damage is considered feasible without the assurance that over time the risk exposure that the program takes on will be reduced through responsible mitigation actions. The standards established by the NFIP are based on a nonstructural approach to floodplain regulation and are designed to supplement the federal government's program of structural flood works. While demand continues for the construction of floodworks to provide protection to floodplain residents, many are still skeptical about the efficacy of such structures. Although the NFIP has adopted a nonstructural approach to floodplain management, the NFIP rate structure gives credit to structural flood works if they are certified to protect against the base flood.

The responsibility for hazard mitigation under the NFIP is split between the federal government and the local participating community. The NFIP enters into an arrangement with the local community whereby structures built in the floodplain without full knowledge of their degree of flood risk can be insured at less than full actuarial rates. In exchange, the local community makes a commitment to regulate the location and design of future floodplain construction in a way that results in increased safety from flood hazards. The federal government has established a series of building and development standards for floodplain construction to serve as minimum requirements for participation in the program. These standards use the 100-year flood as the basis for regulation. The primary mitigation action required by the regulations is elevation of the lowest floor of a structure above the level of the base flood as determined by the Flood Insurance Rate Study and shown on the FIRM. The rates for coverage of structures built or substantially improved after the date of the FIRM are based on this elevation.

The local community is responsible for adopting and enforcing these floodplain management standards, and compliance is accomplished through the building permit process. Since local governments have jurisdiction over land use and development, it is only at the local community level that the implementation of standards will be effective. FEMA, working with the state government, conducts periodic reviews at the local level to assess the local community's enforcement of NFIP standards. A determination that a community is not adequately enforcing local ordinances can result in a period of probation, during which a surcharge is added to insurance premiums on all NFIP policies in the community. If the community fails to take corrective actions during this probation period, it can be suspended from the program, which means that NFIP coverage becomes unavailable.

Periodically, there have been proposals that the federal government be authorized to override local regulation when the local government has refused to participate in the NFIP, thereby denying insurance protection to its citizens, or when a local government does not adequately enforce its floodplain regulations. The NFIP, however, has consistently taken the position that federal land use regulation at the local level is illegal, and, in any case, would be unworkable.

While the NFIP compliance process has identified a number of violations of program standards at the local level, there has never been a comprehensive assessment of the level of compliance nationwide or of the overall effect of program standards on local development patterns. Nevertheless, certain limited assessments have been done and a certain amount of data relevant to the issue has been collected. The Natural Hazards Research and Applications Information Center (NHRAIC) prepared a report in 1992 for the Federal Interagency Floodplain Management Task Force noting a number of significant achievements in floodplain management, including more widespread public recognition of flood hazards, as well as reduced development and losses in many localities. Much of these accomplishments can be attributed to the NFIP. But the report also found that "a considerable distance remains between the status quo and the ideal that can be envisioned" (NHRAIC, 1992). The report acknowledges the difficulty in assessing the effectiveness of floodplain management, stating "that there are few clearly stated, measurable goals, and that there is not enough consistent reliable data about program activities and their impacts to tell how much progress is being made in a given direction" (NHRAIC, 1992).

In 1994, the Interagency Floodplain Management Review Committee (IFMRC) studied the causes and consequences of the 1993 Midwest flooding. The committee's report, while recognizing that NFIP requirements are minimum standards that are applied to areas subject to very different flooding conditions, noted that "in the Midwest, the NFIP tends to discourage floodplain development through the increased costs in meeting floodplain management requirements and the cost of an annual flood insurance premium, although this may not be the case elsewhere in the nation. Individuals and developers appear to choose locations out of the floodplain to avoid these costs. Developers have the added incentive of wanting to avoid marketing flood-prone property. Many communities visited by the Review Committee actively discourage floodplain development" (IFMRC, 1994, p. 97).

An analysis of loss experience from 1978 to 1994 also provides some indications of the effectiveness of the program's standards. The figures show that, in general, structures built before 1975, which were subject to no NFIP standards, suffered about six times more damage from flooding than those built after NFIP mitigation requirements became effective.

Figure 6-2 shows the distribution of NFIP flood claims made from the beginning of the program through December 31, 1997.

Insurance—Rates

Since the NFIP began, its insurance structure has included two general classes of properties: those insured at full actuarial rates and those that, because of their date of construction, are statutorily eligible to be insured at lower, "subsidized" rates—that is, rates that do not reflect the full risk to which a property is exposed. Actuarial rates for coverage are charged to property owners living outside 100-year flood hazard areas and to those living within 100-year areas who built or substantially improved structures after the federal government provided complete risk information from the Flood Insurance Rate Map.

The date of this map differentiates the two classes of properties. Properties built prior to the availability of this information (known as pre-FIRM structures) are insured at "subsidized" rates which, while they represent some contribution on the part of the property owner to his or her financial protection, do not reflect the full risk to which the property is exposed. The 1966 HUD study considered such a "subsidy" to be important in providing an incentive to local communities to participate in the NFIP and regulate future floodplain construction.

The concept of "subsidy" as applied to the NFIP is often misunder-

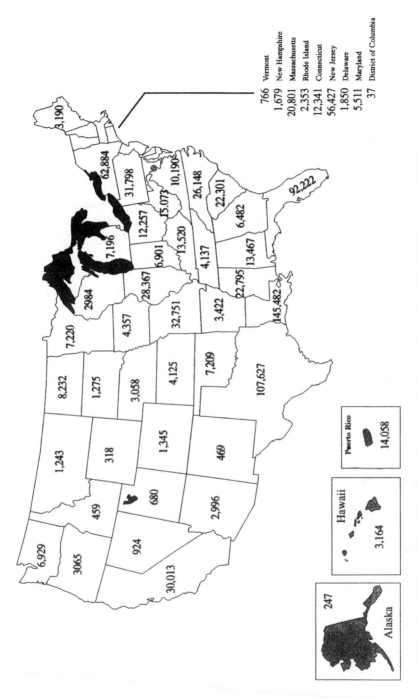

FIGURE 6-2 NFIP total flood insurance claims made through December 31, 1997. Source: FEMA.

stood. The subsidy does *not* refer to a direct infusion of taxpayer dollars to offset the shortfall in premium from properties paying this lower rate. Rather, it prevents the program from building a catastrophe reserve that can be used to pay claims in heavier loss years; the NFIP has been required to exercise its borrowing authority from the U.S. Treasury to pay such claims.

While the law calls for a lower-than-actuarial rate to be charged to pre-FIRM structures in the floodplain, it does not specify the actual rate. During the 1970s, when the primary program objective was to enlist communities to participate in the NFIP, these rates were substantially lower than the actuarial rates. In 1981, the program began a process of gradually increasing the rates charged for existing construction in order to reduce the amount of subsidy and attempt to render the program self-supporting for the average historical loss year. In 1981, rates for pre-FIRM structures were increased 19 percent, and in 1983 they were increased another 45 percent. From 1983 through 1995, eight increases in rates for both structures with subsidized and actuarially based rates averaged about 8 percent each. As the subsidized rates for existing structures have been raised, the number of properties requiring a subsidy has also continued to decline over the years, going from about 75 percent in 1978 to about 35 percent in 1997. The premiums paid by this group of insureds are estimated to be about 38 percent of the full-risk premium needed to fund the long-term expectation for losses.

Insurance—Private Industry Participation

Under the NFIP's original 1968 design, the private insurance sector marketed flood insurance to floodplain occupants through a consortium of 125 insurance companies called the National Flood Insurers Association (NFIA). The NFIA not only sold and serviced flood insurance policies through licensed property insurance agents, it also underwrote the coverage, with the federal government providing premium equalization payments to offset underwriting losses due to the inability to charge full actuarial rates on existing properties in high-hazard areas. This was the program structure recommended in the 1966 study. In 1977, the government, because of problems of financial control and a dispute over program authority, exercised its option under the act to make the program an all-federal program, using private insurance agents and adjusters to sell and service the policies directly through the NFIP. This form of insurance operation prevailed until 1983.

Write-Your-Own Program

In 1983, private insurance companies again became involved in the NFIP through the Write-Your-Own (WYO) Program, an arrangement by which the private insurers sell and service federally underwritten flood insurance policies under their own names, retaining a percentage of premium (currently 31.9 percent) for administrative and production costs. The WYO arrangement, signed every year by participating companies, specifies the responsibilities of both the insurers and the federal government. WYO companies are required to comply with standards outlined in the WYO Financial Control Plan, which covers all aspects of their operations, including underwriting, claims, cash management, and financial reporting. Company representation in the administration and oversight of the program is accomplished through the Institute for Business and Home Safety's Flood Committee, and through company representation on the WYO Standards Committee. In addition, the Flood Insurance Producers National Committee serves as an advisory group to represent the interests of property insurance agents who sell and service NFIP policies.

One hundred sixty-seven companies now participate in the WYO program. Over the 15-year period since WYO's inception, a growing percentage of total program policies have been written by WYO companies. At present, WYO companies write about 92 percent of NFIP policies, the remainder still being written by agents who deal directly with the federal government.

COMMUNITY RATING SYSTEM

The Community Rating System (CRS) was created by the FIA in 1990 as a mechanism for recognizing and encouraging community floodplain management activities that exceed the minimum NFIP standards. The three goals of the CRS are to reduce flood losses, to facilitate accurate insurance rating, and to promote the awareness of flood insurance. When a community's activities meet these three goals, its flood insurance premiums are adjusted to reflect the resulting reduction in flood risk.

In order to recognize community floodplain management activities in this insurance rating system, they must be described, measured, and evaluated. A community receives a CRS classification based upon the scores for its activities. There are ten CRS classes: Class 1 requires the most credit points and results in the greatest premium reduction; Class

10 receives no premium deduction. A community that does not apply for the CRS or does not obtain a minimum number of credit points is a Class 10 community.

Community application for the CRS is voluntary. Any community that is in full compliance with the rules and regulations of the NFIP may apply for a CRS classification better than Class 10. A community applies for a CRS classification by sending completed application worksheets with appropriate documentation that it is doing activities recognized by the CRS. This documentation is sent to the community's regional FEMA office.

A community's CRS classification is assigned on the basis of a field verification of the activities included in its application. These verifications are conducted by the Insurance Services Office, Inc. (ISO), an organization providing standardized forms, and rating and actuarial services to firms in the insurance industry. The ISO has been conducting community grading for fire insurance for many years and is now rating communities under the newly implemented Building Code Effectiveness Grading Schedule discussed below. This organization's available resources enable it to carry out the fieldwork involved with the CRS.

There are currently 912 communities participating in the CRS.

SIDEBAR 6-1 ▬▬▬▬▬▬▬▬▬▬▬▬▬▬▬▬▬▬▬▬▬

Creditable Activities for the CRS

Credit points are based on how well an activity meets the three goals of the CRS mentioned in the text, and are calculated for each activity using formulas and adjustment factors. The CRS recognizes 18 creditable activities organized in four categories:

- *Public information* activities include programs that advise people about flood hazard, flood insurance, and ways to reduce flood damages.
- *Mapping and regulations* activities are intended to provide increased protection to new development. These activities include mapping areas not shown on the Flood Insurance Rate Map, preserving open space, devising higher regulatory standards and storm water management, and preserving the natural and beneficial functions of floodplains. This credit is increased for growing communities.

Tulsa, Oklahoma, and Sanibel Island, Florida, are, at Class 5, the two best-rated CRS communities (25 percent premium discount). Part of the underlying strategy of the CRS is to have communities join and to improve their classifications. In fact, over 32 percent of all CRS communities are a Class 8 or better.

While these 912 CRS communities are only 5 percent of the total NFIP participating communities, they represent over 63 percent of all policyholders. It is important to note that these 912 communities must undertake and demonstrate flood loss mitigation activities above and beyond the significant level of activities already required for minimum community NFIP participation. Even communities that are in Class 10 and in good standing with the NFIP carry out significant flood loss reduction activities. Sidebar 6-1 describes the type of activities that earn CRS credits.

Relationship to Other Insurance Grading Systems

The CRS for flood insurance was inspired by the Public Protection Grading System (PPGS) used by the insurance industry to adjust fire insurance premiums according to a community's firefighting and preven-

---

- *Flood damage reduction* activities involve mitigation programs where existing development is at risk. Credit is awarded for addressing repetitive loss problems, acquisition and retrofitting of flood-prone structures, and maintaining drainage systems.
- *Flood preparedness* activities include programs such as flood warning, levee safety, and dam safety.

Some of these activities may be implemented by the state or by a regional agency rather than at the community level. For example, some states have disclosure laws that are creditable in the flood hazard disclosure category. Any community in those states will receive those credit points if it demonstrates that it effectively implements the law. The CRS recognizes some established methods for obtaining credits in each activity, although communities are invited to propose alternative approaches to these activities in their applications.

tion capability. The grading system approach has been employed and refined for fire insurance since 1912. The expertise and advice of the Insurance Services Office, the industry entity responsible for administering the PPGS, played a valuable part in the development and implementation of the CRS. But whereas the PPGS addresses the engineering aspects of a community's loss reduction capability, the CRS expands the community grading concept to include ordinances and codes that are being enforced by communities and that result in a reduction in flood losses. The insurance industry has developed, and is now in the early stages of implementing, a Building Code Effectiveness Grading Schedule (BCEGS). This system, developed in the wake of the devastating losses caused by Hurricane Andrew, grades communities on their adoption and enforcement of building codes that affect losses from natural hazards for which the private sector is providing insurance. It is safe to say that the successful implementation of the CRS encouraged this expanded community grading concept for use in adjusting private sector insurance rates. This cross-pollination of ideas has been an important and valuable benefit of the private and public sector involvement in the CRS, the NFIP, and other natural hazard mitigation efforts.

PROGRAM FUNDING

The NFIP is funded through the National Flood Insurance Fund, which was established in the U.S. Treasury by the National Flood Insurance Act of 1968. Collected premiums are deposited into the fund, and losses, operating costs, and administrative expenses are paid out of the fund. The National Flood Insurance Program also has the authority to borrow up to $1 billion from the Treasury. During the early years of the program, when the private insurance sector underwrote the coverage, the government called upon its borrowing authority regularly to make the premium equalization payments to the industry pool. This was necessary because no funds had been provided by the act to capitalize the program initially. The NFIP experienced annual operating losses during the period 1972–1980 ranging from $5,400,000 to $323,228,000. Cumulative operating losses from the beginning of the NFIP in 1969 to 1980 totaled about $817,680,000.

Congressional appropriation to repay borrowed funds was first requested in 1981 and the program received an appropriation to repay borrowed funds each year until 1986. The total amount borrowed from the Treasury prior to 1986 was $1.2 billion, all of which was repaid through a series of appropriations. During this same period, from 1981 through 1986, only 1983 and 1984 experienced a negative operating result. Between 1987 and 1996, negative operating results were experienced in 1989, 1990, 1992, 1993, 1995, and 1996.

As noted previously, before 1981 no action was taken regarding the level of subsidy being accorded existing properties in high-risk areas, the program focus being primarily on promoting community participation. Consequently, program expenses inevitably exceeded income. In 1981, the administrator of the FIA established a goal of making the program self-supporting for the average historical loss year by 1988. Note that the goal was that the program be self-supporting, not that it be actuarially sound.

The continuing statutory requirement to "subsidize" certain existing properties in high-hazard areas still makes actuarial soundness an unrealistic goal at this point in the program's history. "Self-supporting" means that in years where losses are less than the historical average, the program builds up a surplus to be used in years when losses exceed that average. However, even though the premiums for existing buildings have been regularly raised to the point where they are now higher, on average, than the premiums paid for new construction, they still represent only about 38 percent of the full-risk premium for these properties. These properties currently make up about 35 percent of the NFIP book of business. This shortfall in premiums prevents the program from accumulating the kind of catastrophe reserves that would reduce the need to borrow. The premium shortfall also makes it more difficult for the NFIP to repay funds when borrowing becomes necessary.

Nevertheless, even with the statutory subsidy, from 1986 until 1995 the program operated with a positive cash balance except for a brief period beginning in December 1993, when the NFIP borrowed $11 million from the Treasury to pay claims. The borrowed funds were repaid in six months. However, in the four fiscal years from 1993 through 1996, the program experienced over $3.4 billion in losses, and beginning in July 1995 the program has had to borrow substantial funds from the Treasury. The level of borrowing reached $917 million in June 1997, but by March 1998, the level had been reduced to $810 million through repayments, including $45 million in interest. The outstanding amount borrowed as of April 30, 1997, was $880 million. It is significant to note that from 1969 through 1997, almost $10 billion in premiums was collected from policyholders, and over $8 billion was expended by the program for loss payments and loss adjustment expense.

Until 1986, federal salaries and expenses, as well as the costs associated with flood mapping and floodplain management, were paid by an annual appropriation from Congress. From 1987 to 1990, however, Congress required the program to pay these expenses out of premium dollars without giving the program the opportunity to adjust the rate structure to reflect these costs. This resulted in a loss of about $350 million in loss reserves available to pay claims. Beginning in 1990, a federal policy fee of $25 (now $30) was levied on most policies in order to generate the funds for salaries, expenses, and mitigation costs.

ISSUES

The remainder of this chapter describes issues confronting the National Flood Insurance Program at this writing, their background and possible solutions.

Interrelationship of Insurance, Risk Assessment, and Mitigation

The key to any success that the NFIP has been able to achieve is the close integration of the three critical program components: risk identification, hazard mitigation, and insurance. The NFIP has been able to integrate these components to some extent, and has been held up as a model for addressing other hazards, such as earthquake, that present some of the same problems of insurability. However, within the current FEMA structure, the risk identification and mitigation responsibilities are organizationally separated from the insurance operation. The two offices have established procedures to assure ongoing coordination of policy decisions, but continued coordination will require a level of effort beyond that needed when the three components were in the same office.

Mandatory Purchase

The 1966 HUD report recognized the importance of the lending industry to the success of a flood insurance program: "The degree to which lenders of all kinds encouraged or required flood insurance as a condition of loans in high-hazard areas would have a great deal to do with its acceptance" (U.S. Department of Housing and Urban Development, 1966, p. 83). The report assumed that lenders would treat flood insurance like other forms of hazard insurance in their mortgage activity. The fact that lenders failed to do so became obvious when Tropical Storm

Agnes struck in June 1972. At the time, there were less than 200,000 flood insurance policies in force (PIF) nationwide, and very few in the areas affected by the storm (29 in Wilkes-Barre, Pennsylvania, one of the worst hit communities). This was one of the major factors that prompted Congress to enact the Flood Disaster Protection Act of 1973, which made participation in the NFIP a prerequisite for eligibility for certain kinds of federal assistance used in floodplains. It made the purchase of flood insurance a mandatory condition of any federally related mortgage on a property in an identified flood hazard area.

Even after the legislative requirement was in place, the spread of coverage to other properties in the nation's floodplains has not been proportionate to the level of mortgage activity in those areas. While the policy count grew significantly in the years immediately following the 1973 act, the annual growth rate since 1980 has averaged only about 3 percent. Florida has 1.5 million PIF, followed by Louisiana's 300,000, Texas's 250,000, California's 235,000, and New Jersey's 155,000. Data on the reasons for this slow growth have been hard to come by. Some evidence indicates that, while lenders have been requiring flood insurance at the point of loan origination, they have not been vigilant in ensuring that the mortgagor renews the policy in subsequent years.

In the light of this evidence, the National Flood Insurance Reform Act of 1994 focused on lender compliance as one of the major program areas requiring attention. The 1994 act expands the mandatory purchase provisions in a number of areas, as follows:

- It requires that lenders who establish escrow accounts for hazard insurance and taxes include flood insurance premiums in such accounts. Allowing the homeowner to pay a portion of the premium each month as part of a mortgage payment eliminates the annual temptation to drop the coverage.
- The act calls for lenders to require flood insurance to be purchased during the life of a loan, whenever it becomes known that the property is located in an identified special flood hazard area. The 1973 act required purchase of insurance only at loan inception.
- In cases where borrowers who are required to purchase flood insurance fail to do so, the act gives lenders the authority to purchase the coverage on behalf of the borrower, that is, to force-place it.
- Unlike the 1973 act, the 1994 Reform Act provides for financial

penalties for noncompliance. The absence of real consequences for lenders not requiring flood insurance had created a situation where the importance of flood coverage as mortgage protection was recognized only at the time of an actual flood, too late to benefit victims.

Mitigation and Existing Structures

A major issue to be dealt with in the area of mitigation is the matter of existing structures. The NFIP floodplain management requirements are primarily directed toward guiding future construction in the floodplain. The 1966 HUD report projected that subsidies for existing high-risk properties would be needed for about 25 years. This was based on the assumption that a turnover in housing stock, particularly those structures most exposed to serious flood hazard, would, over that time, result in the large majority of structures being rebuilt in compliance with NFIP building standards and insured at full actuarial rates. However, modern construction and renovation techniques have greatly extended the useful life of buildings, with the result that the replacement of this older housing stock has not occurred as quickly as anticipated by the feasibility study.

The most serious problem presented by existing flood-prone structures has to do with buildings that are damaged repeatedly but not beyond 50 percent of their value, the threshold that triggers the requirement to rebuild to specified safer elevations. Varying definitions of repeated damaged structures have been used over the years. For purposes of establishing eligibility for increased cost-of-compliance coverage, the 1994 Reform Act defines a repeatedly damaged structure as one "covered by a contract for flood insurance under this title that has incurred flood-related damage on two occasions during a 10-year period ending on the date of the event for which a second claim is made, in which the cost of repair, on the average, equaled or exceeded 25 percent of the value of the structure at the time of each flood event."

This definition, which applies only to properties experiencing 25 percent or more damage at the time of each loss, is more limiting than the one that the program has historically used to track repetitive losses. Properties falling within the 25 percent threshold number about 9,000. If the 25 percent threshold is not used, the number rises to 76,000. However, these properties account for over 200,000 paid losses, almost one-third of the program's total paid losses, and the $2.8 billion paid on these losses represents about 35 percent of the $8 billion paid over the history of the program. One hopeful sign is the fact that less than half of the 76,000 buildings that have experienced multiple losses are currently insured by the NFIP. This suggests that these properties are gradually being removed from the floodplain either by demolition or relocation, consistent with program intent.

Section 1362 of the NFIP legislation was intended to partially address existing floodplain structures. This provision authorized the NFIP to purchase certain insured properties that had been either substantially or repeatedly damaged and transfer the land to a public agency for sound floodplain use. Funds were appropriated for Section 1362 annually from 1980 until 1994, when the Section 1362 program was replaced by the Mitigation Grant Program. Over the period during which funds were available, approximately 1,400 properties were purchased at a total cost of about $51.9 million. Properties were purchased in 28 states, with the largest number in Texas, Illinois, Missouri, Oklahoma, and Mississippi.

While Section 1362 has been of some use in reducing the program subsidy, the 1994 Reform Act focused on the need to provide additional mitigation tools both to protect the financial stability of the National Flood Insurance Fund and to promote a more rapid accomplishment of the floodplain management goals of the program through addressing existing flood-prone structures. The Reform Act's three major provisions were:

1. The establishment of a national Flood Mitigation Fund to provide grants to states and municipalities for a variety of measures to address local circumstances related to serious flood hazard. The funds for these grants come from the federal policy fee and from financial penalties imposed on lenders found to be out of compliance with the mandatory purchase provisions of the act. [The NFIP interpreted the Omnibus Reconciliation Act of 1990 as requiring that the mitigation grant program funds, like other mitigation costs, come from the federal policy fee rather than from premiums (P.L. 101-508).] The Mitigation Grant Program replaces two mitigation programs: the previously described purchase of property under Section 1362 of the act, and the Upton-Jones program (discussed below). It also expands the latitude available to local communities in the use of such funds.

2. Extension of coverage under a flood insurance policy to include the additional cost needed to meet NFIP building standards after

a flood (i.e., increased cost of compliance coverage). Prior to the legislation, the flood policy could only provide coverage to restore a damaged property to its pre-flood condition. This created the awkward situation for the NFIP of requiring that substantially damaged structures meet the program's elevation standards without making available the financial resources necessary for reconstruction. As noted above, the act extended the coverage not only to properties suffering substantial damage (i.e., damage in excess of 50 percent of their value), but also to repetitively damaged properties.

3. Statutory recognition of the Community Rating System. As described previously, the CRS was initiated in 1990 and puts into practice the recommendation of the 1966 report for incentives to states and local governments for wise management of flood-prone areas.

There have been other efforts over the years to address the problem of existing flood-prone structures. In the 1970s, following a severe coastal storm, the NFIP experienced a number of claims on damaged properties in the Outer Banks of North Carolina whose locations made repair and reconstruction on the same site problematic. Ongoing erosion and the structures' locations relative to the shoreline made it virtually inevitable that another storm would damage them again, at further cost to the NFIP. The administrator of the FIA decided that it would be more cost-effective in the long term to authorize an additional expenditure under the policy to move the structures back away from the coast on the same lot rather than simply rebuilding them in the same place.

A later legal opinion by the Office of General Counsel held that such a decision could not be left to the discretion of the administrator unless criteria were established that would make any insured property in a similar circumstance eligible for such a benefit. The use of the policy contract in this manner to promote mitigation objectives generated a certain amount of controversy, and it was decided that it would be too difficult to develop a program that would result in the kind of consensus needed to be effective.

Erosion

One of the major issues that faces the NFIP in coastal states is the problem of erosion, of which the state of North Carolina provides a good illustration.

The North Carolina Division of Coastal Management determined in 1986 that, of the 237 miles of North Carolina's oceanfront shoreline surveyed, approximately 73 percent was eroding, with almost 5,000 buildings inside the 60-year setback line, and 777 inside the ten-year line. The state of North Carolina has an active coastal zone management program under the auspices of the Coastal Resources Commission (CRC), which has designated "areas of environmental concern" (AEC) in which uncontrolled development might cause irreversible damage to property, public health, and the natural environment. One of the four categories of AEC is the ocean hazard system that covers lands that are along the oceanfront and inlets and are vulnerable to storms, flooding, and erosion.

With respect to the erosion hazard specifically, the CRC in 1977 adopted coastal erosion hazard regulations and established statewide oceanfront setback standards. The setbacks are based on average annual long-term erosion rates, natural features at the site, and the type of development. The erosion setback line extends landward from the first line of natural vegetation to a distance equal to 30 times the average annual erosion rate at the site. Large-scale development, such as motels and condominiums, must meet an additional setback requirement.

The current statute governing the NFIP does not allow the rate structure to reflect the erosion hazard, which prevents the program from buttressing local coastal regulation with a financial leverage. The bill that eventually became the 1994 Reform Act originally contained a provision that would have allowed the program to charge for the erosion hazard and to impose setback requirements on erosion-prone properties. This provision was removed from the final version and replaced by a section calling for a study of the economic effects of such a provision.

The 1973 act added flood-related erosion as a covered peril under the NFIP, and in 1985, the Upton-Jones provisions were added to the act, extending coverage to structures whose foundations were being gradually undermined by erosion, but which had not yet suffered any actual physical damage that could qualify as the basis for a claim. Upton-Jones made coverage available under the NFIP so that properties "in imminent danger of collapse" due to flood-related erosion could be either demolished or relocated to a safer site before experiencing a flood.

The effectiveness of the Upton-Jones provisions as a mitigation tool has been questioned. A 1988 Congressional Research Service report to Congress noted that "some critics of the Upton-Jones amendment point out that it does not provide an appropriate *quid pro quo*: the NFIP provides an increased benefit of insurance coverage for erosion-threatened structures; but beneficiaries do not pay the cost of this benefit, either (for policyholders) through increased premiums related to increased risk, or (for communities) through increased mitigation action to prevent future losses" (Simmons, 1988).

The Upton-Jones provision was eliminated by the National Flood Insurance Reform Act of 1994 and subsumed into the new Mitigation Grant Program. During the period when benefits were available under Upton-Jones, 385 structures were demolished, at a cost of $29.9 million, and another 141 structures were relocated, at a cost of $5.2 million. A small number of claims remain to be closed.

Coastal Barrier Resources System

In 1982 Congress enacted the Coastal Barrier Resources Act (CBRA) (P.L. 97-348, 16 *U.S. Code*, sec. 3501 et seq.), which was amended in 1990 by the Coastal Barrier Improvement Act (P.L. 101-591, 104 *Stat.*, 2931). This legislation was implemented as part of a Department of the Interior (DOI) initiative to preserve the ecological integrity of the barrier islands—areas that serve to buffer the U.S. mainland from storms and provide important habitats for fish and wildlife. In order to discourage further development in certain undeveloped portions of the islands, the law prohibits new federal financial assistance, including federal flood insurance, in areas that DOI designates as part of the Coastal Barrier Resources System (CBRS). The CBRS was originally composed of 186 units and was expanded to 560 units in 1990. There are CBRS units in 22 states, Puerto Rico, and the Virgin Islands.

The NFIP implements the coastal barrier legislation by amending the flood insurance rate maps (FIRMs) of affected communities to show the CBRS areas, and issues no policies in those areas. Consistent enforcement, however, is difficult. Since flood insurance applications are processed on the basis of community numbers, and portions of communities having CBRS units eligible for insurance, the NFIP must depend on the vigilance of insurance agents to distinguish which areas of a community are eligible for coverage and which are not.

A review conducted in 1992 by the General Accounting Office (GAO) found not only that significant new development continued to occur in certain CBRS units after the law was enacted, but also that NFIP coverage was written on 9 percent of the residences in the units sampled. Unable to rely on total accuracy of underwriting information at the point of application, the NFIP has depended on post-claims underwriting to deny claims on CBRS properties ineligible for coverage, but the GAO has found this to be a less than ideal response to the problem. The NFIP continues to face a challenge in this area.

Map Revisions

One of the NFIP's important ongoing responsibilities is the updating of FIRMs. These maps need to be updated because of changes in stream channels, local development, coastal erosion, or construction of flood-works. In addition, the 1994 Reform Act required that every community be screened at least once every five years to determine the need for a map revision. Since funds for these revisions are limited, they are prioritized based on a cost-benefit analysis of candidate communities, with the most weight given to communities where development is greatest or most likely.

A particular form of map revision that does not result in the issuance of a new map is the Letter of Map Change (LOMC), which can be either a Letter of Map Amendment (LOMA) or a Letter of Map Revision (LOMR). LOMCs are issued to accommodate properties in special flood hazard areas that are elevated above the base flood level either because of natural topography (in which case the LOMA is issued) or through the use of fill (LOMA-F). In these situations, revisions are generally requested by property owners who want to be exempted from the mandatory purchase of flood insurance. While the LOMA issued because of local topography reflects natural conditions, the LOMA-F based on fill tends to encourage the shift of risk from the elevated property to other properties in the area.

Requests for LOMCs have steadily increased over the past five years. In 1993, there were about 2,800 such requests; in 1997 the number exceeded 11,000. Because the primary purpose for requesting a Letter of Map Change is to excuse the owner from purchasing flood insurance, the entire LOMC process represents a rather shortsighted attitude toward the value of insurance as well as an under appreciation of the flood peril.

Market Penetration

As shown in Table 6-1, NFIP policies in force stand at about 4 million as of January 31, 1998, with total premiums of $1.5 billion and

TABLE 6-1   NFIP Policies, as of January 31, 1998

State Name	Policies	Contracts	Coverage	Premium
Alabama	31,267	23,693	$3,040,828,800	$12,590,206
Alaska	2,269	2,124	253,792,900	894,127
American Samoa	177	147	22,725,800	80,057
Arizona	25,844	25,061	2,787,814,000	9,370,586
Arkansas	13,132	12,994	830,143,900	4,804,696
California	379,227	361,986	54,938,258,200	148,117,394
Colorado	14,601	13,010	1,604,555,200	6,393,360
Connecticut	26,607	22,093	3,422,757,700	15,770,825
Delaware	14,216	10,543	1,766,024,500	6,384,508
District Columbia	335	86	19,704,400	83,424
Florida	1,635,721	1,102,863	190,579,653,100	526,508,110
Georgia	51,823	49,362	6,774,150,700	22,160,783
Guam	155	155	18,201,200	103,000
Hawaii	45,536	14,117	5,334,955,100	13,713,026
Idaho	8,128	8,016	1,228,966,300	2,787,577
Illinois	46,254	42,595	3,995,726,500	19,576,667
Indiana	25,216	25,178	1,731,097,400	10,674,440
Iowa	9,763	9,749	721,136,400	4,616,649
Kansas	10,069	10,032	710,527,200	4,158,461
Kentucky	22,264	21,979	1,350,636,100	8,598,709
Louisiana	334,045	327,369	32,888,330,600	119,883,650
Maine	6,443	6,029	652,649,200	3,524,444
Maryland	45,678	27,843	4,516,394,900	14,800,547
Massachusetts	35,098	31,007	4,162,194,200	21,558,946
Michigan	26,293	25,019	2,241,978,500	11,341,981
Minnesota	13,184	12,889	1,137,582,800	4,572,150
Mississippi	41,378	40,606	3,491,664,200	15,302,304
Missouri	22,754	22,636	1,897,445,800	11,454,638
Montana	9,570	9,397	976,254,500	2,919,701
Nebraska	11,941	11,888	854,533,700	4,776,598
Nevada	11,877	11,791	1,575,761,300	4,993,559
New Hampshire	4,211	3,673	393,918,500	2,097,915
New Jersey	157,534	133,893	19,332,025,400	81,685,889
New Mexico	10,373	10,266	812,552,600	3,824,770
New York	87,543	82,833	10,876,776,400	49,791,682
North Carolina	73,660	65,141	8,843,212,300	31,423,193
North Dakota	12,686	12,497	1,257,150,200	3,949,441
Ohio	33,144	31,864	2,386,721,400	14,329,736
Oklahoma	14,182	14,022	1,076,584,300	5,532,267
Oregon	21,159	20,313	2,675,936,700	9,423,163
Pennsylvania	62,885	61,536	5,590,188,100	29,312,187
Puerto Rico	41,179	36,574	2,118,391,800	12,103,412
Rhode Island	10,579	9,176	1,260,573,200	6,668,231
South Carolina	105,847	74,823	14,796,686,500	43,794,852
South Dakota	4,528	4,528	456,029,100	1,705,060
Tennessee	13,104	12,725	1,262,243,700	5,472,716
Texas	280,215	265,240	32,721,302,200	102,971,224

TABLE 6-1 NFIP Policies, as of January 31, 1998

State Name	Policies	Contracts	Coverage	Premium
Trust Terr. of Pacific	1	1	319,900	496
Utah	2,369	2,015	273,127,800	909,735
Vermont	2,422	2,375	202,402,600	1,247,577
Virgin Islands	2,458	2,169	233,525,300	1,292,508
Virginia	63,200	55,054	7,364,603,500	23,198,088
Washington	25,757	25,093	2,883,529,900	11,054,833
West Virginia	18,072	18,033	1,034,512,200	7,729,741
Wisconsin	11,707	11,584	865,058,800	5,028,062
Wyoming	2,818	2,812	388,351,800	1,092,419
Unknown	58	40	5,992,600	20,756
TOTAL	3,982,556	3,240,537	$454,638,161,900	$1,488,175,076

Source: National Flood Insurance Program.

aggregate coverage of $455 billion. Single-family residences account for almost two-thirds of all policies in force (PIF). Nonresidential (commercial) structures account for only about 4.3 percent of all PIF. Definitive figures on the potential market for flood insurance are difficult to obtain. A conservative estimate is that the current 4 million NFIP policies represent less than half of those property owners who should carry the coverage. The reasons that have been given for this low market penetration are varied:

- Many floodplain residents underestimate the seriousness and likelihood of flooding.
- Lenders have not been especially zealous in requiring the purchase of flood insurance as the law requires.
- NFIP policies are unique, difficult for insurance agents to write, and therefore not marketed aggressively. Because they are single-peril policies, they are viewed as more costly than other lines of coverage.
- Many potential consumers expect that federal disaster assistance will adequately provide for post-flood recovery.
- Many people are simply misinformed about or unaware of the availability of national flood insurance.

The 1994 Reform Act was intended to address some of these factors. The NFIP has also launched a series of initiatives to simplify the process of writing a policy, and to market the product on a scale comparable to other insurance products. While the role of lenders is critical to public acceptance of flood insurance as a routine means of financial protection, there is continuing evidence that floodplain residents need to be better informed about both the availability and the value of flood insurance.

In 1995, the NFIP hired the advertising firm Bozell Worldwide to conduct a major marketing campaign called "Cover America" to raise the national consciousness on the matter of flood insurance and to increase the NFIP policy base. In the first two years of this campaign, the policy base increased by about 20 percent, which compares with an average annual growth rate of about 4 percent.

Subsidy Reduction

Currently, about 35 percent of NFIP policies are subsidized to some degree, which costs the program approximately a half billion dollars in annual premiums. Subsidization was built into the program both to avoid unfairly burdening existing floodplain residents who had built without full knowledge of the risk, and as an incentive to communities to regulate the location and quality of future floodplain construction. To the extent that the mitigation components of the program are effective, the subsidy will necessarily shrink. There have been impressive examples in recent storms demonstrating that mitigation measures that meet program standards are effective in reducing losses. However, the turnover in subsidized housing stock has not occurred as anticipated. Further, a relatively small number of structures that have suffered repetitive damage over the years, but not to a level exceeding 50 percent of their value, have accounted for a disproportionate percentage of the program's total losses. These structures continue to qualify for subsidization, a circumstance addressed by the Reform Act.

Acknowledging the requirement to provide some level of subsidy to existing structures, the program has attempted, since 1981, to accelerate the reduction of the amount of subsidy on individual policies without raising premium costs past the point of reasonableness. The annual rate review continues to result in reduction of subsidies.

Any strategy to reduce the subsidy is influenced by the political climate in which the program has to operate. There are conflicting pressures to both reduce and maintain the subsidy, and the public attitude toward subsidies varies with the national mood at the moment. In the immediate aftermath of a major event like the Midwest flooding of 1993,

*Midwest floods of 1993. Houses protected by levees, St. Genevieve, Missouri
(FEMA, Andrea Booher).*

there is often a general lack of support for allowing people to continue to
occupy the floodplain with no financial consequences. However, when
measures are proposed to address the problem, the public has often for-
gotten the storm and changed its sentiments.

Reducing the program subsidy without losing public support for the
program will depend on remaining attuned to the public mood at the
time. Of course, an elimination of the subsidy that pushes premiums to
levels that are unaffordable for many would also likely increase the demand for alternative forms of public assistance, which in the long run are
more costly than the subsidy. The 1994 Reform Act authorized a study
to assess the economic effects of eliminating the program subsidy. This
study is being conducted by Price Waterhouse and is scheduled for
completion in summer of 1998.

Relationship to Federal Disaster Assistance

The effect of federal disaster assistance on the NFIP is more complex
than may at first be apparent. The availability of federal disaster assistance is often cited as a reason why floodplain residents do not purchase
flood insurance. For example, one of the recommendations of the Inter-

agency Floodplain Management Review Committee, which studied the 1993 Midwest flooding, was that post-disaster benefits to those eligible to buy insurance be reduced. The prevailing public impression is that federal disaster assistance is generally equivalent to the financial protection provided by hazard insurance. In reality this is not the case. In most flood disasters, the primary mode of federal assistance provided to property owners after a disaster is a low-interest loan from the Small Business Administration (SBA). Disaster victims who are deemed unable to repay an SBA loan can receive an Individual and Family Grant (IFG) from FEMA. However, the amount of these grants is limited to a maximum of $13,000, and they are intended to address reasonable needs and necessary expenses, not to make the grant recipient "whole."

Despite the fact that insurance protection is preferable to either a loan or a grant, the public often perceives otherwise, and this public perception serves as a deterrent to the purchase of flood insurance. Lack of interest in coverage is also due to the attitude expressed by many residents of hazard-prone areas that "it can't happen to me," so why think about protection from disasters? In addition, personal budget limitations can prevent many individuals from voluntarily investing in insurance and other mitigation measures (Kunreuther, 1996). The NFIP's marketing campaign has attempted to reveal the relative inadequacy of federal disaster assistance compared with flood insurance and the reality of the risk facing those who live in high-hazard areas.

A less frequent but more problematic form of federal disaster assistance is the buyout program authorized by the Robert T. Stafford Disaster Relief and Emergency Assistance Act. Buyouts are one of the eligible activities of the Hazard Mitigation Grant Program. This program makes available an additional 15 percent of the total federal disaster expenditures in a given disaster to be used for mitigation purposes. The objective of the buyout program following a flood is to use the opportunity presented by a serious flood to pay for the relocation of victims out of the floodplain, thereby reducing the costs of future flood disasters. This program was used extensively after the 1993 Midwest floods, and it was employed in the wake of the 1997 floods in North Dakota, Minnesota, and South Dakota. The city of Grand Forks, North Dakota, was especially hard hit by the 1997 floods and the buyout process is still ongoing for much of the affected area.

When buyouts are authorized, they are available to all affected residents of a targeted area, whether they are insured or not. While the NFIP can accurately claim that insurance coverage is preferable to a disaster loan or grant, a buyout results in a much larger payment to the property owner than an insurance policy. To the extent that floodplain residents begin to presume the availability of buyouts in the event of a flood, the incentive to purchase flood insurance is significantly reduced. Residents of Grand Forks who purchased flood insurance in response to a January 1997 NFIP marketing campaign have voiced this very concern. FEMA is analyzing ways to better coordinate both programs.

The 1973 Flood Disaster Protection Act required recipients of federal disaster assistance who live in the floodplain to purchase flood insurance as a condition of the assistance. The 1994 Reform Act reinforced, and in fact strengthened these requirements by including a provision denying any future assistance to anyone who receives federal disaster assistance in the aftermath of a flood and fails to purchase and maintain flood insurance in force. While this provision is designed to promote the purchase of flood insurance, it has its greatest potential impact on grant recipients, who are generally in lower-income categories than those receiving loans and thus less likely to be able to afford insurance. Whether the threat of denial of future federal assistance will have the intended effect of promoting insurance purchase among this segment of the population remains to be seen. The question must be raised as to whether insurance is the mechanism best suited to provide financial protection to a low-income population.

Future Role of the Private Insurance Industry

As discussed above, the original role contemplated for the insurance industry in the NFIP was to underwrite the coverage, with financial backing as needed from the government. That arrangement ended in 1977, and since that time the industry has performed in various ways in the role of fiscal agent to the program, even with the more extensive involvement of the industry under the WYO program. Initiatives have been proposed to formulate an all-hazards type of insurance that would cast the industry into the role of underwriting the coverage, with provisions for the government to reinsure losses over a certain level. The most recent proposal did not include flood insurance and presumed the continuation of the NFIP. At this point, these initiatives seem to be making little progress. As we gain more and more experience with the flood program, if the subsidy continues to recede, and if the insurers remain committed to the program, it is inevitable that the feasibility of having private insurance companies again underwrite a significant portion of the coverage will surface for discussion. Lloyd's of London, for one, is now able to provide a certain level of private flood coverage because of the more favorable market environment created by the NFIP. One of the concerns expressed over the privatization of flood insurance is whether the vital connection between the availability of flood insurance and the local community enforcement of floodplain management provisions can be maintained. Unless this local enforcement continues, any significant privatization of flood insurance is not likely to be financially feasible.

CHAPTER SEVEN

Mitigation and Insurance

WILLIAM PETAK

NATURAL HAZARD MITIGATION is defined in FEMA's National Mitigation Strategy as "sustained action taken to reduce or eliminate long-term risk to people and their property from hazards and their effects" (FEMA, 1995). Insurance has traditionally served the purpose of reducing the economic impact of individual losses by arranging for the transfer of all or part of the loss to others who share the same risk. Although insurance is not considered a mitigation measure, a carefully designed insurance program can encourage the adoption of loss reduction measures through economic incentives such as premium reductions and lower deductibles.

Mitigating the impacts of natural disasters requires individual or business action and investment to improve a property's ability to withstand the forces of natural disasters. In view of the low frequency of occurrence of catastrophic events, individual homeowners and small business owners tend to be reluctant to allocate scarce resources to mitigation. They understand that the direct benefits from the costs associated with implementing mitigation measures (i.e., return on the investment) are realized only when a natural hazard event occurs and the property experiences reduced damage. Further, the notion of freedom of choice in deciding where to live or locate a home or business and how to design and construct a building remains a driving political force. The cumulative result of many individuals exercising that freedom of choice has been the concentration of many buildings in areas subject to earthquake, flooding, landslide, liquefaction, and severe winds, as well as the design and construction of buildings unable to withstand these natural forces.

One question that needs to be addressed is whether insurance premiums, deductibles, and the availability of coverage are sufficient incentives to cause property owners to voluntarily adopt cost-effective mitigation measures. Some believe that the free market is unlikely to lead to an efficient level of investment in mitigation (Litan, 1991). To be effective in this environment, the role of the insurance industry in facilitating both the voluntary application of mitigation measures and the implementation by state and local governments of mandated loss reduction measures needs to be more carefully defined.

NATURAL HAZARDS AS PUBLIC PROBLEMS

Existing communities and continuing economic development in hazardous areas pose significant risk of catastrophic losses from natural hazard events. Climatic change may also increase the frequency and severity of natural hazard events with catastrophic potential. Although individuals and businesses generally view insurance as the primary tool for reducing the potentially large economic losses associated with natural hazards, insurance is not structured to substantially reduce the impact of a major disaster on the total economic system.

From an insurance company risk-rating perspective, an individual property loss is an independent event involving a small annual risk of loss. Accumulated losses are therefore relatively small, making it possible to manage claims and collect premiums within a high degree of certainty. An individual property loss based upon the traditional rules of insurance would be categorized as a private problem because it would not require large amounts of public aid and the property could be rebuilt with resources provided under the terms of the insurance contract. These single risks, however, when aggregated through the forces of a natural disaster, change from a collection of private, individual problems to a public, community problem; Kunreuther et al. (1978) have characterized the situation as a shift from a private risk to a social risk.

Specifically, the aggregated impacts in terms of homelessness, unemployment, abandonment of property, cost in public services, and so on, are social and economic costs borne by the greater system, and require direct financial assistance from government. Reducing the risk of damage to buildings that serve these social functions is therefore a direct benefit to the taxpaying public following the occurrence of a catastrophic event.

When a private risk becomes a social or public risk, mitigation through both individual (private) voluntary action and government (public) regulatory action is required. Two fundamental issues to be addressed are:

- What factors or incentives are necessary to motivate private property owners to engage in voluntary mitigation actions to reduce the risks from natural hazards, particularly low-probability, high-consequence events?
- Where does the responsibility for mitigation of risks posed by natural hazards shift from being primarily a set of individual private decisions to a public problem requiring public sector intervention (i.e., what is the threshold at which the government must set the primary standards for mitigation of natural hazard risks)?

To address these issues, we need a greater understanding of the risks posed by hazards for both the public and private sectors, the cost and efficacy of various mitigation alternatives, the perceptions of risks by the various stakeholders, and the fundamental factors required to motivate risk reduction behavior. The capacity of local, state, and federal government to address these issues is important. Because private insurance companies are a principal stakeholder in the process, it is also critical that the insurance industry establish broad-based hazard mitigation objectives and programs to guide their own activities and address the problem.

INSURANCE AND MITIGATION

The insurance industry has an established record of supporting the implementation of mitigation measures to reduce risk to property. The industry offered the first model building code for lessening damage from fires, and provided incentives to encourage the adoption and implementation of effective fire hazard mitigation strategies. To emphasize the importance of codes in improving the quality of construction, the National Board of Fire Underwriters stated in their 1949 *National Building*

Code: "It is recognized that good construction is of the utmost importance in every city and town. The competent enforcement of good building law tends to raise the general standard of construction throughout the country. Building codes must provide reasonable safety to the occupants, *must preserve the property itself and must protect adjoining property*" (National Fire Underwriters, 1949, emphasis added).

Although this code set the standard for the insurers' participation in hazard mitigation and for future model codes, it did not cover the risks associated with potentially catastrophic natural hazards. In 1980 the insurance industry, believing "that the model code organizations were ful-

(A) This parking garage at the Northridge Fashion Center withstood the 1994 Northridge earthquake relatively well. It was a cast-in-place structure; note the spirally confined columns. (B) Another Fashion Center parking garage after the earthquake. Although this garage was not as old as the one shown above, it did not fare as well. The inadequate transfer of inertial forces to the shear walls of this precast post-tensional construction may have contributed to its collapse (USGS, M. Celebi).

filling the purpose for which the initial code was established," ceased publication of the National Building Code (American Insurance Association and Alliance of American Insurers, 1995, p. 79).

As pointed out in Chapter 1, the catastrophic disaster losses insurance companies have experienced in the last ten years have sensitized the industry to its vulnerability to natural hazards, and have focused attention on the need for better building construction and hazard avoidance. This experience has caused the industry to review the question of the insurability of natural hazard risks (see Chapter 2) and to become proactive in furthering the cause of better mitigation programs.

To many insurance company officials, the lack of sufficient knowledge about the probabilities associated with a specific natural hazard event, and the uncertainties about the effectiveness of various mitigation approaches in achieving property loss reduction make it difficult for insurers to provide financial incentives for mitigation. The current regulatory system, in which some premiums are fixed below the rates that would be consistent with risk, discourages insurers from adding policyholders to their ranks. If the above conditions prevail, insurers would prefer not to add policyholders by offering premium reductions.

To address the problems of insurability and portfolio risk, both the government and insurance companies rely on analyses performed by consultants who employ various hazard and risk assessment models. Catastrophe models for dealing with earthquake and hurricane hazards are winning increasingly widespread acceptance among insurance firms. However, all the interested parties recognize that in order to link mitigation with insurance it is important to develop models with reduced uncertainty, greater reliability, and improved ability to assess the costs and effectiveness of mitigation alternatives. (For more details on this point see Appendix B, Evaluating Models of Risks from Natural Hazards.)

Legislative Initiatives

One specific outcome of the loss experience in recent years and resulting concerns for business viability was that in 1990 the insurance industry initiated House of Representatives Bill 4480. This legislative proposal was aimed at reducing the exposure of private insurers to earthquakes and earthquake-caused catastrophes by transferring some of the risk to the federal government and by establishing an earthquake mitigation program.

A Federal Emergency Management Agency study of the feasibility of incorporating earthquake loss reduction measures into a federal earthquake insurance program concluded that:

1. There are 15 cost-effective, technically feasible earthquake loss reduction measures, chiefly in land use and building practice, that are acceptable for inclusion in a federal earthquake insurance or reinsurance program.
2. Current earthquake risk analysis techniques, in spite of their uncertainties, are acceptable for the evaluation of loss reduction measures and for the determination of both primary earthquake insurance rates and reinsurance prices.
3. There are two primary vehicles for the effective inclusion of the 15 loss reduction measures into a federal earthquake insurance program:
   — earthquake ordinances for state and local government adoption and enforcement, and
   — a system of risk-based insurance rates that provide financial incentives for the adoption and enforcement of earthquake ordinances.
4. An enhanced federal program of earthquake loss reduction can be justified by the resulting reduction of existing contingent federal liabilities, especially in higher seismic risk zones (FEMA, 1990).

H.R. 4480 was not adopted during the 101st Congress. A new draft legislative proposal with a more clearly stated set of mitigation requirements was prepared by the Earthquake Project of the National Committee on Property Insurance and became H.R. 2806 of the 102nd Congress. However, Hurricane Andrew in 1992 focused the attention of many on the significant potential for catastrophic losses from large hurricane events, and the insurance industry worked for the introduction of House of Representatives Bill 1856 which specifically included the hurricane hazard. The Senate was also considering Senate Bill 1350 related to disaster insurance. S. 1350 was intended to increase the availability of disaster insurance and encourage the adoption of hazard mitigation practices, with the objective of reducing the economic consequences of future natural hazard events and the reliance on federal disaster assistance. This bill would establish three interrelated programs: a residential primary insurance program of earthquake and volcano insurance, a reinsurance program for excess losses experienced by the commercial writers of insurance, and a mitigation program with funding to the states from a surcharge on the primary and reinsurance premiums. H.R. 1856 was similar in its objectives, but it extended the coverage to tsunamis and hurricanes.

Government Agency Reactions to Legislative Proposals

All of the legislative initiatives sought to combine insurance, reinsurance, and hazard mitigation into a unified governmental program. Thus, as with the flood insurance program, these initiatives would have transferred to the federal government a significant portion of the responsibility for insuring the natural hazard risks with the greatest catastrophic potential. Each of the initiatives resulted in congressional hearings. It is useful to consider the concerns expressed by government officials when such far-reaching proposals were being considered.

Frank Reilly, then Deputy Federal Insurance Administrator, in his testimony before the House of Representatives concerning H.R. 4480, stated that any proposal for federal earthquake insurance should (1) account for a correction of a market failure, (2) be based upon actuarial fairness, (3) include sufficient hazard mitigation, (3) have federal oversight and control, (4) be deficit neutral, and (5) be based upon legitimate risk sharing with the private sector.

With regard to hazard mitigation, Reilly stated that the proposal must include provisions requiring participating states and communities to adopt appropriate land use and building code policies that would reduce the risk of loss of life and damage to property. Including these provisions would overcome the "moral hazard" problem of a proposal that shifts a large measure of financial risk to the federal government (Reilly, 1990).

Thomas McCool of the General Accounting Office, after review of S. 1350 and H.R. 1856, stated in testimony before the House of Representatives that (1) setting actuarially sound rates would be a difficult problem because of data and technological limitations, (2) spreading the risk among a large number of policyholders in order to provide affordable rates is difficult to achieve without mandating, (3) making insurance a mandatory purchase as part of federally related mortgages may be ineffective because a significant number of owners do not have federally related mortgages, (4) reinsurance raises many issues about the federal government's exposure to unlimited liability for losses, and (5) many, both in and out of government, recognize that a mega-catastrophe, or a series of smaller disasters in a short time period, could overwhelm the capacity of the industry and result in large payouts by the federal government, thus giving the federal government a significant interest in reducing both the total losses and the federal share of the losses. A well-designed mitigation program with insurance-based incentives for mitigation would help to reduce the risk (U.S. General Accounting Office, 1995).

McCool also noted that, "given the nature of earthquakes, it is difficult to predict probable losses for a given area of exposure." He went on to state that "broad participation is critical to effective risk-sharing and accurate loss predictions" and "depends upon sharing catastrophe risk among a large number of individuals." However, "from a slightly different perspective, if people living in earthquake-prone areas forgo coverage now, homeowners living in low-risk areas may not believe they need the coverage, regardless of how low the premium" (U.S. General Accounting Office, 1995, pp. 6–7).

It is also interesting to observe that in his testimony, McCool stated that even with a mandatory purchase requirement for insurance in the NFIP and premium subsidies, Federal Insurance Administration (FIA) officials found that "only about 20 percent of those living in flood hazard areas have flood insurance" (U.S. General Accounting Office, 1995, p. 4). Thus, flood insurance with its mandatory floodplain management requirements has had limited success in achieving a greater level of mitigation for those living in hazard-prone zones. It appears that as a result of these concerns, the proposed legislation has met with little interest or support in Congress, although the insurance industry continues to work for its passage.

VOLUNTARY AND REGULATORY APPROACHES
TO HAZARD MITIGATION

Mitigation measures for reduction of risk from natural hazards can be achieved through *voluntary action* by property owners, or through the enforcement of a set of *government regulations, ordinances, and codes.*

Voluntary Approaches

Ideally, all property owners would take individual responsibility for evaluating their property and making the necessary adjustments to reduce the risks of damage from natural hazards. In order for voluntary action to occur, property owners need to

- understand the risks associated with natural hazards and the mitigation methods that can be employed to reduce risk to the lowest possible level;

- know the cost and effectiveness of the various mitigation methods; and
- have the appropriate financial incentives and the financial means to take mitigation actions.

It is generally accepted that individual or business property owners will act to reduce the risk of loss from natural hazards when they believe that a significant risk exists and that the risk can be mitigated in a cost-effective manner. Individual property owners must be made aware of the risk, the price of insurance and the degree of attractiveness of specific mitigation measures for their property. If property owners have short time horizons or high discount rates, they will be unwilling to invest in loss reduction measures because they do not perceive the expected benefits to be large enough to justify the expenditure. Those with budget constraints will also be reluctant to incur a certain cost today for uncertain future benefits.

When facilities are considered critical to the continuity of a business operation, professional managers may decide to invest in a hazard mitigation program because it makes good business sense to protect the firm's assets (balance sheet) and shareholder investments from the impact of natural hazard risks. This is of particular concern when a business is publicly owned and management is accountable to the broader financial community through public disclosure rules. The fiduciary responsibility and accountability of corporate financial and operations management will help make the decision to invest in a mitigation program more acceptable.

Regulatory Approaches

Other than voluntary action, the most frequently considered means of reducing natural hazard risks is the application of building codes and land use policies. The implementation and effectiveness of these measures, however, depends on the "institutional capacity" of the local, state, and federal governments (Mittler, 1993). There are four aspects of a government's institutional capacity that permit assumptions to be made about how it will respond to the need for greater mitigation.

First are the statutes and regulations that specify what actions governments are required to take. These typically denote department and agency responsibilities and resources, both human and fiscal, that may be dedicated to mitigation. Second are statutes and regulations that specify relationships between the state and its political subdivisions, cities, and counties. Since disasters are local events, mitigation ultimately depends on the capabilities of the affected area (i.e., trained staff, budget, and political will); however, the assistance states provide, and the laws governing this assistance, will partially determine what eventually gets accomplished.

Third are specific land use management plans and goals. These will most likely include community development plans and codes. Fourth are the experiences of the state and local governments in recent related events. An analysis of what the government did before the event, and what lessons were learned from the event, should provide a guide to future actions.

MITIGATION ISSUES FOR GOVERNMENT AND THE INSURANCE INDUSTRY

Within the complex federal system of government, many issues must be addressed when considering an insurance and mitigation program. First, within the structure of state and local governments, how is the responsibility for mitigation determined? Is there sufficient knowledge to enact and enforce appropriate codes and standards? Can the pressure of special interests be overcome in order to do the right thing? How can insurance programs for low-probability, high-consequence events be regulated so that companies can make affordable coverage available to consumers without risking insolvency?

What role should the federal government play in developing shared responsibility? Can the federal government provide mitigation incentives? What is its role in the education of stakeholder groups? Does disaster relief work against the purchase of insurance or the adoption of loss reduction strategies?

The insurance industry will play an important part in the implementation of a hazard mitigation program, whether cost-effective measures are ultimately adopted through voluntary, individual action or as a result of government regulation. The institutional capacity of the industry— that is, the knowledge, perceptions, motivation, and economic strength of the participating insurance companies—also figures into the mix.

For insurers and other interested parties such as banks and FEMA, one of the most important issues is to determine what role insurance should and can play in adoption and enforcement of appropriate building codes and standards as well as land use practices for both new construction and retrofitting.

The Natural Disaster Loss Reduction Committee of the National Committee on Property Insurance (NCPI) sponsored two building code symposia during 1993. These conferences identified 8 themes (see Sidebar 7-1) that should be central to any industry-based hazard mitigation program, and developed 15 objectives for such a program (NCPI, 1993), all of which are summarized below.

NCPI'S HAZARD MITIGATION PROGRAM OBJECTIVES

Objective 1. Cultural Redefinition of the Purpose of Building Codes. Traditionally, building codes have been designed to assure the basic integrity of structures in order to maintain personal safety. This definition should be enlarged to embrace more comprehensive protection of structures and their contents.

Objective 2. Statewide Codes. Building codes should govern entire states, since statewide codes are easier to comply with, train for, and enforce. They also simplify the work of the design and construction industries, which must deal intensively with their requirements.

Objective 3. Insurer Involvement in Code Development. Insurers should be involved in the process of code development.

Objective 4. A Voluntary Minimum Standard for the Nation. A minimum building code standard should be developed at the national level to serve as a yardstick against which states, counties, and communities could measure their code adoption process.

Objective 5. An Expanded Concept of Prescriptive Codes. The use of prescriptive building codes should be expanded. They should be based on proven performance standards, and should define the construction requirements that must be met in order to achieve a performance level that is specified in the code.

Objective 6. A Two-Tiered Code. A two-tiered code should be developed as an incentive to builders to construct homes exceeding the minimum requirements. The basic code, as it now exists, would be the mandatory minimum standard. An optional, higher-level code would permit builders to note the stronger structural features in their marketing efforts. (The heavy advertising of safety features for automobiles, despite their higher cost, suggests this approach could be successful.)

Objective 7. A Code Enforcement Grading System. A code enforcement grading system, which relates code adoption and enforcement to a community's rate classification, will promote additional code requirements and enforcement.

Objective 8. Enhanced Plan Review Requirements. Minimum standards for plan review should be established and implemented.

Objective 9. Public Education about the Importance of Building Codes for Public Safety. Education programs should be developed that heighten public awareness of the critical role building code departments play in protecting public safety. Funding and staffing of building code departments should also be upgraded so that the departments can meet the demands placed on them.

Objective 10. Retrofitting. More research is needed on retrofitting of existing structures, including research on the costs and benefits of retrofitting. The public should be educated about the value of cost-effective mitigation devices.

SIDEBAR 7-1 ━━━━━━━━━━━━━━━━━━━━━━━━━━━━━━━━━━━━━

NCPI's Central Themes for a Hazard Mitigation Program

1. *Education* is the leading priority. Audiences to be targeted are: the general public; federal, state, and local government; commerce and industry; and academic institutions.
2. *Communication* with other groups is essential. Linked to the education effort, communication and coalition building must occur.
3. *Data collection* can be an integral part of the education process. Reliable statistics are needed (a) to quantify the final cost of a natural disaster; (b) to validate projections developed by simulation models; (c) to assess the cost versus the benefit of retrofit and mitigation methods; and (d) to educate

Objective 11. Public Awareness of Natural Hazard Exposure. The public should be educated about how natural hazards can threaten lives and personal safety, the damage they can cause to property, and the financial costs of such damage.

Objective 12. Professional Education and Training. Education programs should also be directed toward professional, technical, and government/regulator audiences, as well as contractors, job supervisors, and laborers.

Objective 13. Expanded Interdisciplinary Communication. Efforts should be made to promote communication between engineering, design, construction, product design, code enforcement, and insurance professionals.

Objective 14. Premium Incentives. Communities and individuals are more likely to endorse new construction building code enhancements and to employ mitigation measures if encouraged to do so by insurance incentives such as lower premiums.

Objective 15. Develop a "Multi-Hazard" Approach. National and local standards mitigation programs that address all potential hazards must be designed.

---

the public and other interest groups regarding the economic impact of natural disasters.

4. *Dissemination of information* is crucial.
5. *Incentives* and innovative ways of encouraging public support of measures such as retrofitting and mitigation devices, are essential. Two examples using insurance are reduced premiums and/or lower deductibles.
6. *Retrofitting* dwellings and commercial structures should be stressed.
7. *Building codes* must be monitored by the insurance industry on an ongoing basis. The Code Enforcement Grading System is an excellent example of the positive impact insurers can have.
8. *Legislative solutions* could be long-term solutions to the problem of how to deal effectively with the implications of natural disasters.

A SYNTHESIS OF HAZARD MITIGATION
PROGRAM OBJECTIVES

Kunreuther (1996) suggested a program for integrating mitigation with insurance that parallels the objectives arrived at by the NCPI's building code symposia. Kunreuther's program includes six objectives: (1) institute more stringent building codes on new homes, (2) use seals of approval on structures meeting codes, (3) use insurance to encourage hazard mitigation, (4) develop insurance for all natural hazards, (5) institute government reinsurance, and (6) subsidize low-income families.

Given the above objectives and guidelines, the principal means by which the insurance industry can facilitate mitigation investments can be ordered into the following four general categories:

1. *Educate the Public.* The purchase of insurance and the implementation of mitigation actions is highly dependent upon an individual's perception of risk. This perception may have more effect on mitigation actions than any financial incentives. A major role for the insurance industry, therefore, is to conduct public awareness programs to enlighten individual property owners about the risks they face and the mitigation actions they can take to reduce their chances of loss. An informed property owner is assumed to be more likely to engage in risk reduction activities.

2. *Participate in the Model Code Process.* Integrating the interests and needs of the insurance industry with the model code development process and the governmental adoption and enforcement process presents a number of very significant challenges. The industry participated actively in the voluntary building code development process before the 1980s, and it must do so again. At present, even though model building codes contain provisions that can be used to reduce unsafe or dangerous building hazards, they are not generally being used to mitigate the risks to *existing* structures from natural hazards.

    Consider the 1994 Uniform Building Code (UBC), which contains an appendix on the abatement of dangerous buildings, giving discretionary authority to local officials. With the exception of a few cases pertaining to the retrofitting of buildings deemed hazardous when subject to earthquakes, little is being done with respect to existing structures. Through the model code process the insurance industry can make its case regarding the need for better codes to reduce property losses from natural hazards. Insurers and reinsurers have as much at stake in the outcome of these processes as do home builder associations, real estate interests, material suppliers, and local government code officials. After improvement of the model codes, the insurance industry can actively encourage communities to adopt and enforce them.

3. *Provide Financial Incentives.* The most frequently suggested financial incentives are insurance premium reductions, changes in the amount of the deductible, and changes in coinsurance schedules, which reflect the changes in the risk resulting from the implementation of a mitigation program. In the case of premium reductions, individuals will compare the reduction in premium offered with the estimated cost of mitigation action, and decide whether the mitigation measure is beneficial based on a perception of the reduction in risk from its adoption.

   Deductibles and coinsurance involve risk sharing and are designed to encourage the property owner to take actions to protect against small losses; this benefits the insurance company by reducing the expense of dealing with small claims. Thus, a decision to mitigate may not be made solely on an individual's economic rationality. In general, it is assumed that the greater the premium reduction, the more likely it is that the property owner will decide to take mitigation actions; however, the high front-end capital cost associated with mitigation may weaken the financial incentive of a premium reduction spread over many years.

4. *Limit Availability of Insurance.* Property owners will most likely be encouraged to implement mitigation measures if obtaining insurance depends on verification that the property to be insured has been built to an acceptable standard, or retrofitted to reduce the risk of property damage. In addition to existing buildings, all proposed new buildings would need to be designed and built to certain specified codes or performance standards before insurance would be made available. The Federal Insurance Administration has demonstrated the power of conditional availability as an incentive by making flood insurance under the National Flood Insurance Program (NFIP) dependent upon the participating community's adoption and enforcement of floodplain management regulations. Given sufficient market penetration, making the availability of financing and insurance for buildings dependent on their meeting certain high mitigation standards should help motivate builders to build to a higher standard and owners to retrofit existing property.

CONCLUSION

The public increasingly looks to insurance as the mechanism to compensate for losses resulting from many types of risk-taking behavior. Insurance is, however, never an acceptable alternative to loss prevention. In the case of natural hazard risks, it is difficult for most individuals to comprehend the significance of these low-probability, high-consequence risks in their individual lives. The result is inadequate mitigation of natural hazard risks and insufficient insurance coverage. There are no easy explanations, and no easy solutions to the problem of mitigating and insuring against natural hazards. There is, however, an increasing recognition by those in the insurance sector, the model code organizations, and the government that a program must be developed that will address these important issues.

Within the free market perspective, what is needed are changes in the regulatory environment, more imagination by insurers, and incentives for individual businesses and property owners to reduce their own risks. This requires a major change in the way people view disaster insurance and loss reduction measures. Individuals will need to take more responsibility for their risk-taking behavior. Property protection, and thereby the maintenance of community economic viability, needs to be a primary focus of building codes and land use regulations.

Insurers cannot assume that the existence of codes assures the preservation of property and thus need to advocate their importance (Cheshire, 1990). For example, state insurance regulators can restrict the sale of natural hazard insurance to properties built to the standards contained in certain building codes; or as in the case of the National Flood Insurance Program, Congress can require local communities to adopt floodplain regulations in order to qualify for flood coverage. The federal government can, when it determines that there has been a market failure, enter the market and provide an insurance-based mitigation program for all natural hazards.

In summary, there is significant potential for linking insurance to the adoption of mitigation measures that reduce damages to existing buildings and limit the amount of new development in hazardous areas. Making the linkage is very important if individuals, businesses, and government are to benefit from reduced losses.

CHAPTER EIGHT

Regulation and Catastrophe Insurance

ROBERT KLEIN

T HE PUBLIC HAS A VESTED interest in insurance. Hence, public choice plays a significant role in the financing and control of risk. Insurance companies, products, and markets are closely regulated, primarily by the states. Laws and public institutions significantly influence how society responds to a variety of perils that cause human and economic losses. This is especially true for natural hazards, where current private and public institutions (e.g., the legislature; the courts; federal, state, and local agencies; etc.) are struggling to accommodate the greatly increased risk of catastrophes caused by earthquakes, hurricanes, and other acts of nature. We need to understand how current regulatory policies are dealing with this crisis and consider more efficient and equitable approaches for mitigating and financing catastrophic losses.

Current insurance regulatory policies are driven by economic, political, ideological, and bureaucratic forces (see Klein, 1995; and Meier, 1988). The economic rationale for insurance regulation is based primarily on the market failures that are caused by imperfect information and by principal-agent problems associated with the fiduciary aspect of insurance contracts. In theory, the regulators' job is to limit the insurers' risk of insolvency and ensure "fair" market practices. Public policy is not forged in a political vacuum, however, and various factors have affected the scope and design of insurance regulation. Some of these factors include voters' perceptions and preferences regarding how the cost of risk should be shared among different groups.

Insurance regulation affects society's response to the risk of natural hazards in a number of ways. Regulation influences the supply and purchase of disaster insurance by controlling insurers' entry into and exit from the market, their capitalization, investments, diversification of risk, prices, products, underwriting selection, and trade practices. Government coverage requirements and information programs also affect individuals' and firms' demands for disaster insurance and hazard mitigation. Regulators are faced with the difficult challenge of assuring an adequate supply of "affordable" insurance coverage at a time when many insurers are seeking to significantly decrease their disaster risk exposure and increase their prices. The resolution of this dilemma could have substantial implications for the economies of many disaster-prone areas and the individuals who live in those areas.

This chapter assesses the role that insurance regulation has played in society's response to disaster risk, and identifies the issues that must be addressed in evaluating how regulatory policy might be modified to help lower disaster costs and support a more efficient and equitable way to finance them. The first section provides a brief overview of the structure of insurance regulation, focusing on its primary institutions and functions. This is followed by a detailed evaluation of the areas of regulation that are most relevant to insuring disaster risk and the options available to regulators to improve market conditions.

The chapter concludes with a discussion of how insurance regulation could be better coordinated with other government policies to achieve public goals. As noted in Chapter 1, insurance regulation interacts with many other areas of public policy that affect catastrophe risk. Hence, the implications of insurance regulatory policies for insurance market conditions could vary depending on the nature of other public and private sector actions that affect catastrophe risk.

STRUCTURE OF INSURANCE REGULATION

The current state regulatory framework for insurance is rooted in the early 1800s when insurance markets were generally confined to a particular community. (For reviews of the history of insurance regulation, see Hanson et. al., 1974; Lilly, 1976; and Meier, 1988.) Local stock companies and mutual protection associations formed to provide fire insurance to property owners in a city. The high concentration of risk and large conflagrations led to highly cyclical pricing and periodic shakeouts when a number of insurers failed after a major fire (Hanson et al., 1974). The local orientation of insurance markets at the time led municipal and state governments to establish the initial regulatory mechanisms for insurance companies and agents.

Government control of insurers was initially accomplished through special legislative charters and discriminatory taxation, but these proved to be inefficient mechanisms as the number of companies grew and the need for ongoing oversight became apparent (Meier, 1988). New Hampshire appointed the first insurance commissioner in 1851. Insurance commissions were subsequently formed by various states to license companies and agents, regulate policy forms, set reserve requirements, police insurers' investments, and administer financial reporting. Price regulation in the early 1900s was essentially confined to limited oversight of property-casualty industry rate cartels. Through the years, insurance department responsibilities grew in scope and complexity as the industry evolved. Price regulation also changed as industry rating bureaus were transformed into advisory organizations and insurance pricing became very competitive.

Several major forces appear to have heavily influenced the evolution of insurance regulatory functions and institutions. One factor has been the dramatic growth and increasing diversity of insurance products and the types of risks that insurers have assumed. Increased competition among insurers and alternative risk-financing mechanisms have imposed additional pressures on the industry and its regulators. Competitive pressures and consumer demands prompted insurers to incur increased financial risk during the 1980s, which resulted in a significant increase in insurer failures and guaranty fund costs (Klein, 1995). This increased risk has been coupled with the geographic extension of insurance markets nationally and internationally, which has increased the interdependence among regulatory jurisdictions.

The relationship between the federal and state governments has been a critical factor in the evolution of insurance regulation and also is very pertinent to disaster insurance issues. Throughout the industry's development, the states and the federal government have wrestled over its regulation. This tension is fed by the interstate operation of many insurers and their significant presence in the economy. On numerous occasions, the federal government has sought to exert greater control over the insurance industry. The states have fought back aggressively to retain their authority, backed by the industry. (See Meier, 1988, and Advisory Commission on Intergovernmental Relations, 1992, for reviews of various attempted federal interventions into insurance regulation.)

The primacy of the states' authority over insurance was essentially affirmed in various court decisions until the Southeastern Underwriters case in 1944 [*United States v. South-Eastern Underwriters Ass'n*, 322 U.S. 533 (1944)]. In that case, the U.S. Supreme Court ruled that the commerce clause of the U.S. Constitution applied to insurance and that the industry was subject to federal antitrust law. This decision prompted the states and the industry to join forces behind the passage of the McCarran-Ferguson Act in 1945, which delegated regulation of insurance to the states, except in instances where federal law specifically supersedes state law.

The federal government has affected state insurance regulatory policy and institutions in several ways. In a number of instances, Congress has instituted federal control over certain insurance markets or aspects of insurers' operations that were previously delegated to the states (e.g., the provision of liability insurance by risk retention groups, employer-sponsored health plans, etc.). In other cases, the federal government has established insurance programs, such as flood and crop insurance, that are exempt from state regulatory oversight. In addition, the federal government has set regulatory standards that the states are expected to enforce (e.g., Medicare supplement insurance). Finally, federal policies in a number of other areas such as antitrust, international trade, law enforcement, taxation and expenditures, and the regulation of banks and securities have significant implications for the insurance industry and state regulation. See Figure 8-1 for a schematic representation of the parties involved in the regulatory process. Arrows are not drawn between the boxes because all of the parties are connected with each other in some ways.

Authority of State Insurance Commissioners

State insurance commissioners have the principal authority to regulate the business of insurance within their jurisdictions. This authority encompasses: licensing insurers and producers; oversight of insurers' financial structure, investments, transactions, and operations; approval of insurance products and prices; and policing insurer and producer trade practices. The insurance commissioner is authorized to take direct control of an insurer if its financial condition becomes hazardous to policyholders' interests. Insurance commissioners' authority is broad and can have significant effects on insurance markets.

At the same time, insurance regulatory agencies are part of a larger governmental structure that helps to set and implement policy as well as constrain regulatory actions. Regulatory policy is formulated collectively by the insurance commissioner, the legislature, and the courts. The state legislature establishes the insurance department, enacts insurance laws, approves the regulatory budget, and may review and approve insurance regulations. Insurance departments are part of the state executive branch, either as a stand-alone agency or as a division within a larger department. Commissioners must often utilize the courts to help enforce regulatory actions, and the courts in turn may restrict regulatory action.

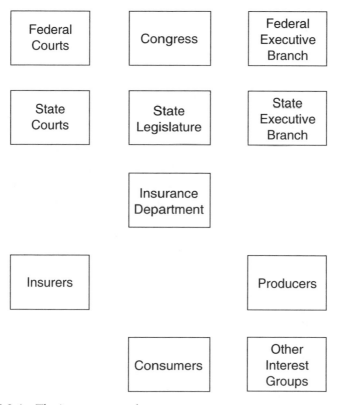

FIGURE 8-1 The insurance regulatory system.

Each insurance department also must coordinate with other state insurance departments in regulating multistate insurers, and must rely on the National Association of Insurance Commissioners (NAIC) for advice as well as some support services. As discussed above, the federal government overlays this entire structure, delegating most regulatory responsibilities to the states, while retaining an oversight role and intervening in specific areas. (In practice, the federal government has left the principal insurance regulatory functions to the states.)

Most commissioners are appointed by the governor (or by a regulatory commission) for a set term or "at will," subject to legislative confirmation. Typically, the governor and other higher administration officials do not interfere with daily regulatory decisions, but may influence general regulatory policies and become involved in particularly salient issues. Twelve states elect their insurance commissioners, who are autonomous in the sense that they do not take orders from the governor, but they must still cooperate with the administration and legislature in order to achieve their objectives. Elected commissioners directly cultivate voters' political support, while appointed commissioners do this indirectly as part of their administrations.

Principal Regulatory Functions

Insurance regulatory functions can be divided into two primary categories: solvency regulation and market regulation. Solvency regulation seeks to protect policyholders and the public against the risk that insurers will not be able to meet their financial obligations. Market regulation attempts to ensure fair and reasonable insurance prices, products, and trade practices. Solvency and market regulation are closely related and must be coordinated to achieve their specific objectives. Regulation of rates and market practices will affect insurers' financial performance, and solvency regulation constrains the prices and products that insurers can reasonably offer.

All U.S. insurers are licensed in at least one state and are subject to solvency and market regulation in their state of domicile and in other states in which they are licensed to sell insurance. If an insurance company is licensed in a state, it is classified as an *admitted insurer*; an insurer who operates in a state but is not licensed is classified as a *nonadmitted insurer*. Reinsurers domiciled in the United States also are subject to the solvency regulation of their domiciliary state. Some U.S. and non-U.S. insurers write certain specialty and high-risk coverages on a nonadmitted or surplus lines basis that are not subject to price and product regulation. Insurance commissioners still control entry of nonadmitted insurers into their states by imposing minimum solvency and trust requirements, and regulate surplus lines brokers.

Solvency Regulation

The public interest argument for the regulation of insurer solvency derives from inefficiencies created by costly information and principal-agent problems (Munch and Smallwood, 1981).[1] If an insurance company becomes insolvent, any unfunded obligations to its policyholders do not become the personal liability of the insurance company's owners. (This is the case with virtually all organizational structures used by insurers; the only exception is Lloyd's organizations.) Hence, owners of insurance companies are not necessarily motivated to maintain a high level of financial safety because their personal assets are not at risk to cover these obligations.

It is costly for consumers to properly assess an insurer's financial strength in relation to its prices and quality of service. Consequently, consumers may unwittingly purchase a policy from a financially weak or high-risk insurer. This diminishes an insurer's incentive to limit its financial risk if its behavior is not rewarded by increased demand and consumers' willingness to pay a higher price for its services. Many insurers may have other reasons to limit their financial risk. For instance, many insurers have a significant amount of franchise value (i.e., investments that will yield long-run profits if the insurer stays in business) that will be lost if the insurer becomes insolvent. However, some insurers may be motivated to take advantage of consumers' ignorance and incur a high probability of ruin in return for higher profits.

Principal-agent conflicts also arise after insureds have purchased policies from insurers. It is difficult for policyholders to monitor and control the behavior of their insurers once the policyholders have paid premiums in return for the insurers' promise to pay claims covered under their contracts. Hence, some insurers may be encouraged to increase their financial risk beyond a level acceptable to their policyholders and jeop-

---

[1]This point is well recognized in the economic literature on financial institutions (see Klein, 1995). For example, Transit Casualty Insurance Company effectively gambled policyholders' funds by growing rapidly, incurring excessive risk, and charging inadequate prices, all of which ultimately led to its failure.

ardize their ability to meet future claims obligations. Insurers can increase their financial risk in many ways, including investing in high-risk assets, underpricing to attract new business, establishing inadequate loss reserves, failing to adhere to appropriate underwriting guidelines, failing to purchase adequate reinsurance, paying excessive dividends to owners, and incurring an excessive exposure to catastrophic losses. These actions can increase short-run profits and returns to owners of an insurer at the expense of the insurer's long-run solidity and its capacity to meet its obligations to policyholders.

Without regulation, such actions taken by insurers could result in an excessive number of insolvencies. This is not to say that all or even most insurers would choose to incur excessive risk if left on their own. However, experience has indicated that some will. For example, nine insurers failed due to losses from Hurricane Andrew because of their relatively low levels of capital and their failure to take adequate precautions to protect themselves against catastrophic losses. It is this kind of insurer that regulators must identify and control. Solvency regulation is intended to impose an upward bound on the degree of insolvency risk in accordance with society's preference for safety.

Regulators protect policyholders and society in general against excessive insurer insolvency risk by requiring insurers to meet certain financial standards and to act prudently in managing their affairs. State statutes require insurers to meet minimum capital and surplus standards and financial reporting requirements, and authorize regulators to examine insurers and take other actions to protect policyholders' interests. Solvency regulation polices various aspects of insurers' operations, including capitalization, pricing and products, investments, reinsurance, reserves, asset-liability matching, transactions with affiliates, and management.

If an insurer fails to comply with regulatory requirements and/or its financial condition becomes hazardous to policyholders' interests, regulators are authorized to intervene in various ways to control the insurer's activities. If an insurer becomes severely impaired or insolvent, regulators will place it in receivership and either rehabilitate, sell, or dissolve the company. State guaranty funds, financed by uniform pro rata assessments on the premiums of solvent insurers, cover most of the unfunded obligations of failed insurers to policyholders and claimants. While guaranty funds provide a level of financial security for policyholders and claimants, they also create a moral hazard problem that must be controlled by effective solvency oversight by regulators.

Market Regulation

Market regulation encompasses a diverse set of areas and is approached somewhat differently by the various states. Rates and policy forms are subject to some form of regulatory approval in virtually all states. The traditional explanation for regulation of insurance prices also involves costly information and solvency concerns (Hanson et. al., 1974; Joskow, 1973). Some insurers may charge inadequate prices to grab a larger volume of business and increase their short-run profits at the expense of their long-run financial solidity. Some consumers will buy insurance from low-price carriers without properly considering the greater financial risk involved. According to this theory, poor incentives for safety could induce a wave of "destructive competition" in which all insurers are compelled to cut their prices below costs to retain their market position. Thus, regulators must impose a floor under prices to prevent the market from disintegrating because of a downward spiral of price cutting and insolvencies.

The desire to make sure premiums were high enough against particular risks governed insurance rate regulation until the 1960s, when states began to disapprove of requested price increases in lines such as personal auto and workers' compensation. The rationale offered for restricting insurance price increases is that the costs consumers incur in shopping for insurance products impede competition and lead to excessive prices and profits by insurers (Harrington, 1992). Further, imperfect consumer information and unequal bargaining power between insurers and consumers can expose potential policyholders to abusive marketing and claims practices of insurers and their agents.

It also can be argued that it is costly for insurers to ascertain consumers' risk characteristics accurately, giving an informational advantage to insurers already entrenched in a market and creating barriers to entry that diminish competition and result in much higher prices than would be expected based on the insured risk (Cummins and Danzon, 1991). In this view, the objective of regulation is to enforce a ceiling that will prevent prices from rising above a competitive level and to protect consumers against unfair market practices.

In addition, the public may express a preference for regulatory policies to guarantee certain market outcomes consistent with social norms or objectives. For example, most states require drivers to purchase auto liability insurance, and this increases political pressure to suppress auto insurance prices to keep them affordable for all drivers. Similarly, the necessity for homeowners' insurance may increase public pressure on regulators to keep homeowners' rates low, even in areas subject to high catastrophe risk.

State laws typically require that rates *not* be inadequate, excessive, or unfairly discriminatory. For the personal property-casualty lines, approximately half of the states require rates to be approved before they go into effect. Other states allow insurers to implement rates on personal lines without prior approval, placing greater reliance on competition to regulate prices. Except for workers' compensation and medical malpractice, commercial property-casualty lines in most states are also subject to a competitive rating approach. Under such a system, regulators typically retain authority to disapprove proposed rate increases if they find that competition is not working, although in practice such a finding rarely occurs. [This residual authority is provided in the NAIC's Property and Casualty Model Rating Law (File and Use Version), which many states have adopted. NAIC publishes its model laws in a compilation *Model Laws: Regulation and Guidelines*, which can be obtained from the NAIC.]

Historically, after the passage of the McCarran-Ferguson Act, many property-casualty insurers adopted rates filed by an advisory organization (e.g., Insurance Services Office) or filed deviations from advisory rates. Recently, the states have gone to a system in which advisory organizations file *prospective loss costs*. These loss costs represent *expected* claims payments and associated claims adjustment expenses. Historical loss costs are developed and trended forward to develop prospective loss costs applicable to the policy period for which the advisory loss costs are filed.[2] The advisory loss costs include a provision for catastrophe losses, which are modeled because historical experience alone provides an insufficient basis upon which to estimate future catastrophe losses.

Insurers are allowed to file multipliers to advisory loss costs, which include provisions for adjustments to the advisory loss costs, expenses, profit and investment income, or full rates as before. For example, an insurer can accept the advisory loss costs and file a multiplier of 1.4 to

---

[2]Loss development and trending are standard actuarial procedures to estimate expected or prospective loss costs for the policy period that rates are intended to cover. Loss development is the estimation of the full incurred loss costs from prior years, which include actual claims payments plus estimates of losses incurred but not yet paid. Trending involves projecting historical loss costs forward for a future period based on how loss costs are expected to change over time due to inflation and other factors such as changes in the frequency of claims.

"load" the advisory loss costs for its expenses and profits. In this instance, the insurers' rates would be equal to the advisory loss costs multiplied by the factor of 1.4. Additionally, the insurer could also adjust its multiplier upward (or downward) on the basis that its losses will be higher (or lower) than the advisory loss costs. Alternatively, an insurer might choose to file full rates rather than a multiplier to be applied to the advisory loss costs.

The loss cost approach to advisory filings is intended to promote independence and competition among insurers by removing full advisory rates as a potential focal point for insurer pricing. It is presumed that loss costs will tend to be similar among insurers and less subject to discretion. (Insurer differences and discretion with respect to expected loss costs should only arise from differences in underwriting stringency, risk portfolios, and loss control efforts.) The analysis of industry loss costs by advisory organizations is perceived as having informational value for insurers and the public that outweighs any negative effect it may have on competition. On the other hand, insurers' loadings for expenses and profits are perceived as subject to greater discretion based on insurers' service levels, efficiency, and competitive considerations. Hence, the advisory filing of a common loading factor or full rates would be perceived as having a greater detrimental effect on competition than any efficiencies that would be gained.

In addition, insurers must obtain approval for the products that they sell and, specifically, the policy forms that they use. Regulators seek to ensure that policy provisions are reasonable and fair and do not contain major gaps in coverage that might be misunderstood by consumers, or that might be inconsistent with state coverage requirements. For example, regulators would be unlikely to approve homeowners' insurance policies that would exclude coverage for windstorm losses unless these losses are covered by another mechanism.

Regulators also police insurers' and agents' sales and underwriting activities to make sure that they adhere to certain standards and that claims are handled according to the provisions of insurance contracts. The objective is to prevent abusive practices—for example, misleading presentations on insurance coverage or the failure to pay legitimate claims on a timely basis—that take unfair advantage of consumers. Responding to consumer complaints and performing market conduct examinations are the primary ways in which insurance departments regulate market practices.

State insurance departments perform various other functions. The

insurance commissioner is generally held accountable for the overall performance of insurance markets under his or her jurisdiction, and this
leads to a number of activities to support market operations. Enhancing
consumer information about insurers' prices, products, and financial
strength is a critical function, given the heavy reliance on competition to
ensure good market performance. Other functions performed by insurance departments include agent licensing and education, adjuster licensing (in some states), anti-fraud enforcement, coordinating market assistance plans, collecting premium taxes, and providing public information.
In addition, insurance commissioners frequently advise legislators on
laws affecting public policy on insurance. For example, an insurance commissioner may take an active role in promoting programs to mitigate
catastrophe losses.

Regulatory Constraints

Insurance is a large and complex business; state regulators face significant challenges in achieving their objectives in the face of limited information. Regulators must establish and implement rules governing insurance transactions based on incomplete knowledge about the nature of
the risk transferred and its cost. At best, regulators can attempt to enforce general parameters on market activity and prevent the most fla-
grant abuses without creating significant market distortions inconsistent
with public objectives. This necessitates considerable reliance on market
forces, even in jurisdictions with extensive regulatory restrictions.

The more regulators seek to control insurance market transactions,
the greater the demands on their human and technological resources.
The cost of regulation is further increased by its administration at the
state level, which limits the economies of scale that individual insurance
departments can achieve (Grace and Phillips, 1996). Maintaining a regulatory infrastructure—which consists, for example, of investments in databases and information systems—involves certain fixed costs, as well as
variable costs that increase less than proportionately with the number of
insurance transactions regulated. These costs could be spread over a
larger volume of business if insurance regulation were consolidated into
one or several agencies nationwide. Another example of the inefficiency
of state insurance regulation is the fact that an insurer must make separate rates and forms filings in every state in which it does business, and
regulators in each state must review these rates and filings, thus duplicat-
ing the review process that regulators in other states are performing.

Regulation imposes further indirect costs on insurers in complying with regulatory requirements.

The states also face the challenge of coordinating their actions and reconciling their economic interests in the oversight of multistate and multinational insurers. On average, 80 percent of the premiums in a state are written by insurers that are domiciled elsewhere. Hence the commissioner must rely on these other states to regulate nonadmitted companies' actions with respect to potential insolvency (Klein, 1995). A state focuses its regulatory oversight on the solvency of its domestic companies and the market activities of all insurers operating within its boundaries.

Some analysts contend that many insurance departments lack the expert staff and technology necessary to perform their regulatory responsibilities (see U.S. GAO, 1979). These resources are particularly relevant to the regulation of disaster insurance markets, which requires utilization of considerable scientific and actuarial expertise. Many insurance departments do not possess an abundance of staff with these skills, although they can contract with outside consultants to analyze rate filings. Still, departments face budget constraints that limit the level of analysis they can perform.

Department staff levels vary significantly depending on the size of the markets they regulate and other factors (NAIC, 1997). In 1996, the number of state insurance department personnel, excluding the four U.S. territories, ranged from 23 in South Dakota to 1,056 in California. Total full-time equivalent staff for all departments combined amounted to 9,649, in addition to 1,661 contract staff. For fiscal year 1998, state department budgets ranged from $1.2 million in South Dakota to $122.4 million in California, with a total combined budget for all departments of approximately $753.7 million. Insurance departments employed a total of 90 property and liability actuaries and 354 property and liability rates/forms analysts (not counting contract staff). Many of these analytical resources are concentrated in the larger states, which coincidentally possess a large portion of the risk of natural disasters.

In an effort to overcome some of their jurisdictional boundaries and diseconomies, the states have utilized the NAIC extensively to coordinate their efforts and pool their resources to improve the efficiency of insurance regulation. The NAIC is a private, non-profit association of the chief insurance regulatory officials of the 50 states, the District of Columbia, and the four territories. It was established in 1871 to coordinate the supervision of multistate companies within a state regulatory framework, with special emphasis on insurers' financial condition.

The NAIC functions in an advisory capacity and as a service provider for state insurance departments. Its objective is to allow states to focus their resources on regulation of their markets and the solvency of their domiciliary companies, utilizing support services from the NAIC. The NAIC has several committees and working groups that evaluate disaster insurance issues. These efforts help to overcome some of the structural challenges that the states face in regulating a global insurance industry.

REGULATION AND DISASTER RISK: ISSUES AND OPTIONS

The market problems and market failures with respect to catastrophe risk and insurance pose a number of issues for regulation. To what extent can regulators correct or offset market failures to improve economic efficiency and equity in the provision of catastrophe insurance? What limitations do regulators face? Are there some market problems that would be more effectively addressed by private or public mechanisms outside of regulation? Can regulation potentially worsen rather than improve market conditions? In reviewing the potential benefits and pitfalls of insurance regulation, we need to thoroughly understand the nature of the problems that it intends to correct and whether regulators can reasonably be expected to correct these problems, given economic and political realities.

Regulation and Its Effect on Availability and Price

Regulators are subject to various limitations in seeking to influence the availability and price of insurance in response to economic and political pressures. If regulators suppress rates below costs, insurers will be disinclined to write insurance policies voluntarily. This will diminish the availability of coverage. Regulatory restrictions on policy terminations may delay insurers' withdrawal from a market, but they also may discourage entry and decrease the supply of insurance in the long run. Regulators pressured to constrain price increases and cutbacks in coverage in the short run face the prospect of causing more severe market dislocations in the long run.

Policymakers may seek to justify market restrictions on the basis of equity considerations (e.g., equalizing rates between low- and high-risk areas), but such efforts are likely to impair market efficiency by distorting sellers' and buyers' decisions. Moreover, there are different opinions on whether uniform rates are more equitable than actuarially fair rates that vary with the level of risk.

An individual state's ability to ensure adequate insurance coverage of catastrophe perils and allocate the cost of this coverage among different risks also is limited by its jurisdictional boundaries. States are limited to pooling catastrophe risks within their boundaries. A state cannot create an insurance pool that includes exposures outside its boundaries and force insureds in other jurisdictions to subsidize its insureds.

However, the division of the principal solvency and market regulation responsibilities between states can lead to some inconsistencies between these policies and create externalities between states. This externality problem arises from the fact that the gains that result from regulatory actions that increase the risk of an insurer's insolvency can be distributed in a different manner among states than the costs arising from the insolvency would be distributed. The costs of an insurer's insolvency are distributed among states proportionate to the insurer's volume of business, but the gains from high-risk regulatory actions accrue only to those states that engage in these actions. For example, some states with a high exposure to natural disasters may seek to externalize some of their costs or risk to other states by suppressing insurers' rates or attempts to reduce their exposures. If an insurer becomes insolvent because of these actions, the insolvency costs will be distributed among all states in which the insurer does business, but only the states that suppressed the insurer's rates and underwriting will achieve any gains.

Competition and regulation in other jurisdictions and federal budgetary restrictions check this behavior, but these checks are imperfect, particularly in the short run. For instance, if an insurer would attempt to raise its rates above its expected costs in Iowa to offset its regulation-suppressed rates in Florida, other companies could take business away from the insurer in Iowa by charging lower rates that would not contain a subsidy for Florida insureds. The federal government also may limit the disaster assistance it will provide to a state. However, as explained above, a state can increase the insolvency risk of an insurer and externalize a portion of this increased risk to other states. States on the losing end of this practice can only discourage it by exerting moral suasion on the "offending" regulators and forcing insurers to structurally insulate their operations and surplus from claims from high-risk states. It is unlikely that such drastic action would be taken until an insurer's vulnerability became very apparent and forced regulators to respond.

Control of Market Entry and Exit

Regulatory policy and issues in this area vary, depending on whether the insurer is admitted (licensed in the state), nonadmitted (unlicensed), or alien (domiciled in a foreign country).

Admitted Insurers

The states control the market entry of insurers primarily to limit insolvency risk and enforce the regulation of insurers' market practices. Capital requirements for licensing are relatively low, but other aspects of the processing of licensing applications can significantly slow entry. Entry barriers that do not protect consumers impede competition unnecessarily and restrict the supply of insurance. Easing entry barriers could help new insurers to enter catastrophe insurance markets and write business shed by insurers which have an excessive geographic concentration of catastrophe exposures.

Easier entry could be facilitated through greater standardization of procedures and interstate cooperation in admitting nondomestic insurers. Indeed, some states, such as Hawaii, have explored the use of the NAIC's accreditation program as a means to expedite the admission of insurers domiciled in certified states. Reciprocal agreements among states in the admission and regulation of nondomestic companies also could reduce entry barriers. Several states are currently developing standardized and expedited licensing procedures that consider where an insurer is already domiciled and licensed. At the same time, states with severe availability problems should guard against a tendency to set their entry and solvency requirements too low, which could have adverse consequences for consumers in the long run.

In recent years, regulators also have attempted to restrict exit to some degree, particularly with respect to the personal lines and catastrophe risks. Exit refers to insurers' efforts to shed some exposures within a state, as well as efforts to completely withdraw from one or more lines within a state. Regulators may attempt to delay or prevent insurers' withdrawal from a market to maintain the availability of insurance coverage, but these efforts could prove to be counterproductive in the long run if they discourage entry. Exit can be costly for an insurer to the extent that it will lose its investments in establishing a distribution system, as well as its reputation and its relationships with its policyholders. Indeed, the costs of full exit from a state by a multiline, multistate company tend to make it a permanent decision and discourage insurers from such exits.

This gives regulators some leverage in seeking to keep insurers in a market, but there is a limit to regulators' power in this regard.

At most, regulators can slow and increase the cost of exit but not ultimately prevent it. Many states today impose some restrictions on exits in the form of prior notice requirements on policy cancellations and nonrenewals. Insurers serving multistate commercial risks will have a higher cost of exit than insurers that can exit a state without adversely affecting their ability to cover risks in other states; this increases the economic rent that a state can extract before it forces an insurer out of the state.

If insurers attempt to shed exposures after a catastrophe, regulators can impose moratoriums on policy terminations, such as in Florida after Hurricane Andrew, or force an insurer to exit from all lines, not just lines chosen by the insurer. (Florida extended the moratorium in 1996 to business that insurers placed on their books in anticipation of the moratorium being lifted.) Such moratoriums can discourage insurers from writing new business because the moratoriums limit the insurers' flexibility in shedding business if it proves to be necessary to reduce their concentration of risk exposure.

It is possible that exit restrictions could force an insurer into insolvency, but in practice this has rarely occurred. In most cases, insurers, with the help of the courts, can exit the market before they become insolvent. Of course, this is less likely in the event of a catastrophe, which can cause a rapid change in an insurer's financial condition before it is able to withdraw. In extreme cases, insurers may respond to severe exit barriers by establishing a single-state company to insulate the rest of its business from losses within a state.

Some regulators believe there is a need for regulatory restrictions on the speed of insurers' withdrawal or adjustment of their concentration of exposures. This presumes that some insurers "overreact" to perceived increased catastrophe risk and withdraw from a market "too quickly," resulting in an excessive degree of market instability. This could be true, but it has not been demonstrated by empirical analysis. What principles should govern regulatory exit restrictions and how far should they go? At what point would the social cost of exit restrictions exceed the social gains from promoting greater market stability? As an example of exit restriction, a one-year limitation on policy terminations in high-risk areas following a catastrophe could give insurers and regulators the opportunity to take stock of the situation and determine what long-run adjustments are necessary in terms of insurers' concentration of exposures.

Such adjustments, based on sound risk analysis, would allow the market to reach a sustainable equilibrium without attempting to permanently anchor insurers to an excessive concentration of exposures in high-risk areas. In the interim, regulators and consumers could make provisions for alternative sources of coverage.

Without further analysis, it is not possible to determine whether this policy would be more efficient than placing less severe restrictions on insurers' ability to terminate policies after a disaster. Short-term restrictions are more likely to increase efficiency if: (1) insurers' decision-making processes are flawed to the extent that they overreact to external and internal pressures to avoid risk after a disaster; and (2) abrupt decreases in the supply of insurance impose significant transaction costs on insureds, who must seek other sources of coverage or retain more risk than they would prefer. However, long-term or extended moratoriums on policy termination are likely to be counterproductive in terms of their negative effect on entry and on insurers' willingness to write new policies.

Even if regulatory exit restrictions provide some stability to the market in the short run, they can lead to increased market concentration and a decreased supply of coverage in the long run. Exit restrictions also can adversely affect the policy objective of limiting insurers' insolvency risk. Indeed, regulatory constraints on entry and exit could interfere with the market's movement toward an optimal level of risk diversification among insurers. Ideally, as some insurers reduce their excessive concentration of exposures in high-risk areas, other insurers would see opportunities and come in to fill the gap. This process will not guarantee "affordable" rates but it should decrease the "risk premium" that insurers require to cover catastrophe perils.

Consequently, in the long run, regulatory restrictions on the movement of insurers and their capital could have the perverse effect of decreasing availability and increasing prices. If regulators place severe, long-term restrictions on exits, new insurers will be discouraged from entering the market because of fear of the high cost of exit if they need to withdraw. At the same time, insurers already present in the state will find ways to gradually decrease their exposures (even if it is costly to do so) and require higher rates to compensate for the excessive level of risk they are forced to retain. Alternatively, regulators may wish to explore ways in which insurable risks might be more smoothly transferred from insurers with high concentrations of catastrophe exposures to insurers with low concentrations. As shown in Chapter 5, after Hurricane Andrew, Florida forced insurers to renew most of their homeowners' policies. The state has also been forging agreements with insurers to encourage risk transfer.

Nonadmitted and Alien Insurers

Nonadmitted and alien insurers present a special issue. Today these insurers tend to fill gaps in regular markets and cover risks eschewed by admitted insurers. While regulators impose only minimal restrictions on nonadmitted carriers, some barriers still exist, and they could be evaluated in terms of their impact on the availability of disaster-related coverages. Although the role of nonadmitted insurers is understood and accepted in certain commercial lines, what potential role do they have, if any, to provide additional capacity in the personal lines in catastrophe-prone areas? Would it be necessary and feasible to increase regulatory protections for personal risks purchasing coverage from nonadmitted insurers? Is it possible to allow personal risks to form groups or cooperatives to purchase personal lines coverage from nonadmitted carriers? The use of groups would presumably enhance consumers' bargaining power and ability to protect themselves from transactions adverse to their interests. However, some regulators are concerned that group policies and group marketing can be used to discriminate against risks considered to be undesirable by insurers. Also, policymakers would need to consider the problems created by the insolvency of nonadmitted carriers insuring personal lines risks that would not be protected by guaranty funds.

Regulatory restrictions hamper the ability of alien insurers to enter state insurance markets on an admitted basis. To do so, these non-U.S. companies must acquire a U.S. company, establish a U.S. branch, or utilize a particular state as a port of entry with quasi-domiciliary regulatory responsibilities. Within the structure of state insurance regulation, certain requirements are necessary to enable regulators to enforce alien insurers' compliance with state laws and regulations. It might be possible to establish cooperative arrangements among states and foreign jurisdictions to ease the entry of non-U.S. insurers on an admitted basis to serve primary commercial and personal lines markets while preserving necessary regulatory protections. But are alien insurers interested in serving U.S. markets on an admitted basis? And is it possible to allow them to do so, yet maintain a level playing field with respect to their competition with U.S. insurers? Some alien insurers could have advantages in terms of broader geographic diversification, access to capital and reinsurance, and expertise in insuring catastrophe risk. Alien insurers could be a significant source of additional catastrophe insurance if regulatory barriers were eased, but they also could leverage their entry into other lines of insurance which might create economic problems for U.S. insurers. Increased competition from alien insurers could increase financial risk for some U.S. insurers or force them out of business.

Alternative Insurance Mechanisms

The states and the federal government also control the formation and use of "alternative risk-financing" mechanisms such as self-insurance, captives, and risk retention groups. What potential role do these mechanisms have in addressing disaster risk and how should they be regulated? Presumably, such mechanisms would only be functional for catastrophe risk if they allow for risk-spreading across different types of risk, different geographic areas, and over time. The nature of catastrophe risk—infrequent, severe, and concentrated losses—may pose problems for alternative mechanisms that could not efficiently diversify this risk. For example, a reciprocal insurance arrangement for a group of homeowners living in close proximity to each other might work well for auto liability risk, but not for earthquake or hurricane risk. On the other hand, a firm that owns a number of establishments dispersed throughout the country might be in a better position to self-insure, or use a captive to finance its catastrophe risk exposure. Alternative mechanisms may be the most viable for groups and firms that can diversify catastrophe risk within the group or firm, take advantage of favorable tax treatment of accumulated reserves, and/or buy excess reinsurance.

It has been suggested that banks' entry into insurance could ease market availability problems. If banks' insurance activities are confined to distribution, entry costs could be lowered for insurers seeking to write more business in underserved markets subject to catastrophe risk. If banks' activities are expanded to insurance underwriting, this could add capacity to catastrophe insurance markets. However, changes in state and federal law would be required to allow banks to underwrite insurance risk, and this would raise a host of logistical, policy, and regulatory issues. Underwriting disaster insurance is risky and requires considerable expertise. Hence, it may be the least suitable line for bank entry. Also, the correlation of a bank's exposure to insurance losses and loan defaults stemming from a natural disaster would further concentrate its risk. In reality, banks may have little interest in assuming the catastrophe risk exposures that insurers are seeking to shed.

Financial Regulation

Financial regulation and the other areas of regulation are closely tied. In general, more stringent financial regulation means that insurers are forced to offer less favorable terms on insurance contracts (i.e., insurers have to charge higher prices or reduce coverage). On the other hand, regulatory attempts to force insurers to offer more favorable terms to consumers (e.g., lower prices, more generous coverage, etc.) can have a negative effect on solvency. For example, higher capital requirements may force an insurer to raise its rates; regulatory suppression of rates below costs can threaten an insurer's solvency. Thus, financial regulatory policy can affect disaster insurance market conditions, and regulation of disaster insurance markets can affect insurers' solvency.

Capital Requirements

There is an inherent tension between the regulatory objectives of limiting insolvency risk and ensuring available and affordable coverage. Regulators would generally agree that solvency is their first priority, but the availability and affordability of insurance coverage also are important objectives, and these objectives must be balanced consistent with market realities and societal preferences. For instance, regulators could virtually ensure that no insurer would ever fail, by imposing very stringent capital requirements and other risk limitations. However, such a policy may decrease the supply of insurance and raise prices to socially unacceptable levels. In practice, regulators and the marketplace tolerate some degree of insolvency risk in return for a greater supply of insurance. Guaranty funds have been established to diversify this residual insolvency risk across all policyholders, but the "flat pricing" of insolvency guarantees can impose cross-subsidies and encourage excessive risk-taking by some insurers. Regulators could force a suboptimal solvency-supply solution inconsistent with social preferences.

This tension between the goals of solvency of insurers and available, affordable coverage for consumers is exacerbated by the division of primary regulatory responsibilities for regulating a multistate insurer's solvency and market practices between domiciliary and non-domiciliary states. Under "normal" market circumstances, when coverage is readily available and premiums are relatively affordable, a non-domiciliary state may regulate the solvency of a nondomestic insurer more stringently than would a domiciliary state. This is because the domiciliary state has the most to gain from the growth of the company in terms of employment and tax revenues, while the risk of insolvency is distributed among all the states in which the company writes business (Klein, 1995). However, when a state is suffering from severe availability problems and cost pressures, it may favor less stringent solvency regulation of its nondomestic insurers to increase the supply of insurance and relieve pressure on its market.

For instance, a high-risk state may prefer that other states ease catastrophe risk limitations on its nondomestic companies to avoid a significant number of policy terminations. While such a measure might increase the long-term insolvency risk of these insurers, the regulators in a high-risk state may give more weight to the short-term political gains to be had from maintaining the availability of insurance coverage. If an insurer does become insolvent because of its excessive concentration of catastrophe exposures in high-risk states, the current guaranty fund system distributes the insolvency costs among all of the states in which the insurer does business. Hence, all of the other states bear a portion of insolvency risk from catastrophes in the high-risk state, but do not reap the corresponding short-term political benefits from maintaining the availability of coverage. The potential difference in interests between states with respect to an insurer's catastrophe risk could lead to conflicts between the states in how they regulate the insurer.

Areas related to solvency regulation that are likely to have the greatest significance for disaster insurance market conditions include capital requirements, investments, reserves, risk limitation, and reinsurance. As mentioned above, state capital requirements are relatively low and, hence, would not be expected to have a significant effect on the supply of disaster insurance. The states have recently implemented risk-based capital (RBC) requirements that vary according to an insurer's volume of business and other risk factors. While RBC requirements are typically higher than state-fixed minimum capital requirements, RBC requirements are still relatively low (very few insurers fail to meet them) and do not consider catastrophe risk (Cummins et al., 1995; and Barth, 1996). Hence, the introduction of RBC has likely had a minimal effect on entry or insurers' risk of insolvency due to catastrophic losses. For example, all nine insurers that failed because of Hurricane Andrew would have satisfied NAIC RBC requirements prior to their insolvency had they been in effect.

Other Financial Regulation

Regulatory requirements governing investments generally address proper valuation of investments, credit quality, and diversification.

Stricter restrictions on insurers' investments generally lower their investment return and hence require higher insurance prices or some other offsetting action. However, some regulatory restrictions might also hamper insurers' ability to diversify risk, particularly regulation of insurers' use of derivative instruments to hedge financial and underwriting risk, including disaster risk. Because of the complexity and relative newness of these instruments, regulatory policy in this area is still evolving. The NAIC's Investments of Insurers Model Act allows insurers to use derivatives to engage in hedging transactions and certain income-generation transactions. The total value of these transactions is limited to certain percentages of an insurer's total admitted assets. Thus, in theory, the model law should allow insurers to purchase securities that would compensate them for a portion of their losses caused by a natural disaster. The model law does not appear to allow insurers to purchase catastrophe derivatives for income-generation purchases, however.

This prohibition could impede the diversification of disaster risk within the industry. Some insurers might be well positioned to assume more disaster risk because they do not already have high catastrophe exposures. They may find that assuming catastrophe risk through investing in catastrophe-related securities would be less costly than entering a new disaster insurance market as a primary writer or as a reinsurer. Insurers may be more inclined to buy catastrophe-related securities than other investors because of the insurers' greater understanding of the underlying risk.

However, regulators may find it difficult to ascertain whether an insurer's purchase of disaster-related derivatives exposes it to excessive risk. Regulators can choose to: (1) allow insurers considerable discretion in making such investments; (2) expend considerable resources in analyzing and approving specific investments; or (3) establish arbitrary rules that severely limit or prohibit such investments. Some regulators might elect the third option as the easiest and safest route, but this would be contrary to the interests of disaster-prone states. Arbitrary restrictions on insurers' purchase of disaster-related securities could have a negative effect on the ability of financial markets to help diversify disaster risk and expand the supply of insurance.

The issue of risk limitation received increased attention at the NAIC after recent natural disasters. Regulatory action to force an insurer to reduce its catastrophe exposure could have a negative effect on the availability of insurance in disaster-prone states. Such regulatory actions would compel insurers to terminate policies and decline applications in high-risk areas, which would reduce the supply and availability of insurance in these areas. Rating agencies, such as A. M. Best and Standard & Poor's, also have begun to consider an insurer's catastrophe risk in their rating evaluation, which increases insurers' incentives to diminish this exposure.

The valuation of reserves is another area that could affect disaster insurance markets. Currently, generally accepted accounting principles do not allow insurers to establish catastrophe reserves. There is no provision or requirement for catastrophe reserves under statutory accounting rules, nor do insurers receive favorable tax treatment of catastrophe reserves under current law. Reserves for future catastrophes represent a new concept in that they would not be tied to an event that has occurred, but rather would be based on an event that might occur in the future (Davidson, 1996; Russell and Jaffee, 1995). Insurers would be encouraged to accumulate funds to pay future catastrophe losses if these reserves were either not taxed or benefited from tax deferral. Insurers and regulators are currently working on proposed regulations governing catastrophe reserves which would be implemented if the federal government gives catastrophe reserves favorable tax treatment.

The complexity of catastrophe risk assessment, and the uncertainty surrounding estimates of the risk, also would affect estimating proper reserves for future catastrophe losses; insurers and regulators could easily make different subjective judgments about what is appropriate. If regulators require an insurer to increase its reserves for reported or anticipated claims due to natural disasters, this could force the insurer to reduce its exposure or increase rates. Since reserves are reported as a liability, insurers must set aside additional assets equal to the increase in reserves to maintain a given level of surplus. Further, because of the uncertainty surrounding reserve estimates, insurers must set aside additional capital to provide for losses that might exceed reserves.[3] Insurers could raise additional capital to support these reserves, but this would come at a cost that would need to be reflected in insurance premiums. Otherwise, insurers will be forced to reduce their exposure to losses to bring it into line with regulatory requirements. This would necessitate

---

[3]This assumes that insurers have not adequately estimated their catastrophe exposure and already set aside appropriate catastrophe reserves, nor have they set aside additional capital to cover a shortfall in their reserves in the event of a catastrophe. The NAIC's risk-based capital formula would raise insurers' minimum capital requirement if insurers increased their reserves to cover catastrophic losses, unless these reserves were excluded from the formula.

policy terminations and declinations, which in turn would reduce the supply and availability of insurance.

Finally, regulatory requirements governing insurers' ability to claim credit for reinsurance could limit the supply of disaster insurance. Regulators require reinsurers to meet certain criteria to be considered "authorized" and allow a ceding insurer to account for anticipated recoveries from these reinsurers as assets. Primary insurers may still have some incentive to transfer disaster risk to unauthorized reinsurers (i.e., reinsurers who do not meet regulatory requirements to be authorized), but the inability to claim surplus credit for this reinsurance is a substantial disincentive for such transactions. Hence, regulation of reinsurance credit could have a substantial effect on primary insurers' capacity and ability to insure disaster-prone areas. This could hinder the emergence of new reinsurance capacity to help alleviate disaster insurance availability problems if regulators are reluctant to authorize new reinsurers.

The process by which regulators authorize reinsurers and allow reinsurance credit could be evaluated to see if there would be ways to boost capacity and still satisfy regulatory objectives. The access of U.S. regulators to alien reinsurers' financial data and regulatory actions has been one of the issues of contention. Agreements between U.S. regulators and foreign jurisdictions on information disclosure and the disposition of assets in the event of financial difficulty could help more alien reinsurers to become authorized.

Guaranty Funds

The use of guaranty funds and regulatory actions against failing insurers are important components of solvency regulation. Guaranty funds can accommodate short-term capacity problems by boosting assessment limits and borrowing funds, but high insolvency costs from a large natural disaster could severely tax the capacity of a guaranty fund and the market supporting it. As pointed out in Chapter 5, the nine insolvencies caused by Hurricane Andrew exceeded the Florida guaranty fund's assessment limits and forced it to borrow funds against future assessments to pay claims. The current system of uniform, pro rata guaranty fund assessments is not sensitive to risk and, hence, encourages greater risk-taking by insurers and policyholders.

Risk-sensitive assessments would encourage safer behavior, but they also could further tighten the supply of disaster insurance. Still, this could be preferable to arbitrary limits on risk. There also may be some value in exploring arrangements that would allow state guaranty funds to pool capacity in certain instances (e.g., natural disasters) to provide additional liquidity and further diversify insolvency risk. Such arrangements would need to be structured to avoid perverse incentives and externalities among states in their regulatory policies towards high-risk insurers and markets (Klein, 1996). Without such provisions, a state could be encouraged to expose insurers operating within its jurisdiction to an excessive level of insolvency risk, since the costs of insolvencies would be spread among all states.

Policy Forms and Coverage Requirements

In regulating insurers' policy provisions covering disaster risk, insurance commissioners must balance the objectives of available coverage and affordable rates. Historically, regulators have tended to frown on policy provisions that would significantly restrict coverage, because of their concern that insureds will retain an excessive exposure to risk, knowingly or unknowingly. There may be some value in reconsidering this view for catastrophe insurance and allowing insurers to modify coverage provisions (e.g., increasing deductibles, altering limits, establishing coinsurance provisions, etc.) to encourage loss prevention and make it more economically feasible to offer catastrophe coverage. For example, the Insurance Services Office (ISO) has developed new homeowners' insurance policy forms that raise deductibles for damages caused by windstorms. Allowing such policies could serve the public interest if it enabled policyholders to bear more of the cost of small losses themselves, but to retain protection against catastrophic losses that would exhaust their assets. This would help to lower insurers' total losses from a disaster but would still provide insurance protection for severe damage to individual properties. Examining distributions of claim severity related to disasters could help insurers and regulators assess the potential gains from such an approach.

The question of whether homeowners should be mandated to carry certain coverages is difficult to answer. Mandatory coverage requirements can facilitate geographic risk diversification within a state and possibly counter adverse selection. Such requirements can force homeowners in low-risk areas to contribute actuarially fair premiums to a pool of funds available to pay losses in any part of the state. Adverse selection is countered by forcing low-risk homeowners to purchase coverage when they might otherwise choose not to buy insurance. However, the fairness and effectiveness of such requirements are questionable, particularly if regulators also attempt to enforce cross-subsidies through price regulation or other measures. A preferable alternative may be to emphasize consumer education about disaster risk and to avoid policies that encourage risk-taking and discourage the purchase of insurance.

Rates

Regulation of disaster insurance rates is one of the most controversial issues facing policymakers. If an insurance market is structurally competitive, rate regulation should be unnecessary, unless the objective is to enforce cross-subsidies. Enforcing cross-subsidies through price regulation is difficult to accomplish if the offer and purchase of insurance are voluntary. Insurers will be reluctant to offer coverage at prices below cost, and consumers will be less inclined to buy coverage for which the price exceeds the actuarial cost. Some insurers would be encouraged to only market insurance in low-risk areas where they could undercut the prices charged by insurers forced to subsidize their insureds in high-risk areas. (This occurred in Michigan when the legislature attempted to restrict rate differentials between urban and nonurban areas for auto and homeowners' insurance.) Regulatory efforts to suppress rates below costs will cause economic distortions and potentially threaten insurers' solvency if exit is restricted. Figure 8-2 depicts states that have prior-approval regulations for homeowners' insurance rates as well as those states that have instituted regulations restricting rates for coverage against catastrophic losses.

However, allowing insurers to significantly raise prices following a natural disaster raises some economic questions as well as political problems. Recent disasters appear to have awakened insurers to their greater catastrophe risk and the inadequate catastrophe loads in their rates. New databases and more sophisticated modeling techniques have been developed to meet insurers' information needs and support more accurate pricing. However, this science is still evolving, and regulators as well as insurers are having some difficulty assessing the validity of the new catastrophe rate analyses and the assumptions that drive their results. For proprietary reasons, modelers are reluctant to disclose their assumptions in public rate proceedings. The complexity of this analysis, the secrecy surrounding it, and the significance of the results compound the political problems. Hence, insurers have been frustrated in their attempts to ob-

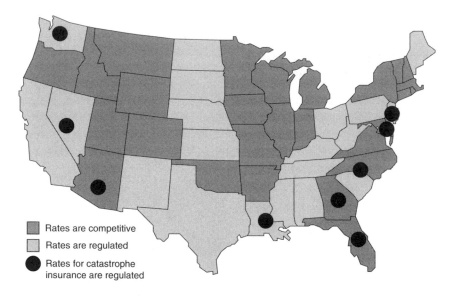

FIGURE 8-2 Insurance price regulation, by state.

tain regulatory approval for large rate increases for increased catastrophe risk based on catastrophe models.

The challenge of prospective pricing of catastrophe risk may require new regulatory approaches. As discussed in Chapter 5, the Florida Commission on Hurricane Loss Projection Methodology is an example of a public-private effort to improve the science underlying catastrophe risk assessment and insurance pricing, and reach consensus regarding them. This approach does not guarantee regulatory acceptance of a catastrophe model; for example, Florida Insurance Commissioner Bill Nelson has challenged a model approved by the commission. The NAIC is developing a manual to help regulators understand and assess catastrophe models. Given the high degree of uncertainty associated with catastrophe loss projections, it is critical to find ways to increase public confidence in the use of the best scientific methods to support adequate insurance prices. There is also the question of whether prior regulatory approval of complex catastrophe rate filings is feasible and necessary. Would it be more practical for regulators to implement some form of competitive rating approach (e.g., file and use), where they would monitor market pricing and performance and intervene only if competition failed or other serious market problems developed?

Discounting premiums for policyholders who take effective mitigation measures is another important issue. Insurers allege that inadequate regulatory ceilings on prices discourage them from implementing pricing discounts for loss prevention efforts by insureds. Their logic may be that any economic gains that would accrue to an insurer from offering such discounts would be offset by greater losses from writing more business at inadequate rates. Research aimed at better catastrophe risk assessment as well as the effects of mitigation on reducing losses could help to solve both problems. If insurance companies and regulators can agree on adequate overall rate levels, insurers should offer price discounts that, at the margin, are equal to their expected cost savings from mitigation. This would increase homeowners' incentives to invest in mitigation and lower overall losses to the extent that it is cost-effective to do so through mitigation. Ideally, a homeowner should receive sufficient information to capitalize investments in mitigation over the expected life of their home. Capitalized mitigation investments could be recaptured in the price of a home if the owner sold it. The present value of even a small annual premium reduction could be significant if accumulated over the life of a home.

Even with discounts for mitigation and greater public confidence in catastrophe risk modeling, implementing actuarially fair disaster insurance prices could have significant economic implications for some high-risk areas and could encounter significant political opposition. Large rate increases could strain some homeowners' budgets and reduce property values in high-risk areas. While these kinds of economic adjustments are inevitable unless subsidies are provided, policymakers may wish to consider phase-in or transition strategies that will help to ease the adjustment process and diminish political opposition.

Regulators and legislators face strong political pressures to subsidize insurance rates in high-risk areas. Political support for subsidies, which may be concentrated among residents in high-risk areas, conflict with countervailing political pressure against subsidies in low-risk areas. Equity issues and political considerations aside, however, it is very difficult for regulators to enforce cross-subsidies within a state as well as across states when insurance is sold by private firms (as opposed to a government-imposed monopoly insurer). Even with coverage mandates, individual private insurers can find a number of ways to avoid writing risks at rates below cost (e.g., not marketing to high-risk areas). Efforts to enforce cross-subsidies also encourage greater risk-taking and economic losses. If rates are suppressed below expected costs from catastrophes, insureds will have diminished incentives to mitigate their risk in order to lower their premiums.

Underwriting Selection

States generally impose only limited underwriting restrictions on insurers, such as requirement of advance notice for policy terminations. However, as indicated in Figure 8-3, a number of states have imposed special underwriting restrictions related to catastrophe risk, according to an NAIC survey. These restrictions have taken the form of limitations or moratoriums on insurers' ability to terminate or otherwise not renew policies because of geographic location or exposure to catastrophe risk.

In addition, in areas subject to severe availability problems, legislators may be tempted to enact laws that restrict insurers' ability to refuse to accept new risks. The Florida legislature, for example, in 1996 considered but did not enact legislation that would force an insurer to maintain a market share in coastal areas at least as high as its market share in other parts of the state. Such restrictions are difficult to enforce and

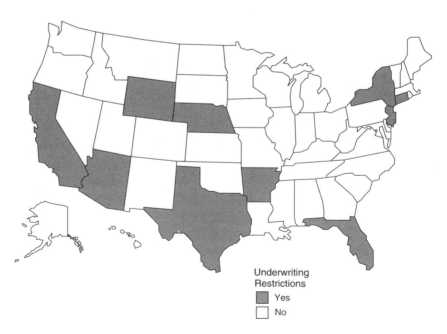

FIGURE 8-3 Catastrophe underwriting restrictions, by state.

counterproductive in the long run. Many of the variables that affect an insurer's volume of business, such as competition, are not under the direct control of the insurer or of regulators. Also, this type of regulation would discourage the entry of new insurers.

As an alternative, insurers could be required to inform rejected insurance applicants as to the reason for their failure to obtain a policy and give the applicant the opportunity to cure the problem that caused the rejection. This might introduce a greater degree of discipline and accountability in the underwriting process without tying insurers' hands, although it would increase paperwork burdens for insurers. This policy could also encourage homeowners to undertake certain mitigation efforts to make their properties more insurable. On the other hand, such a policy would not solve the problem for homes rejected because of their locations or other conditions for which there is no effective remedy.

Claims Adjustment

Regulation of insurers' claims adjustment practices following a disaster can have a significant effect on the cost and supply of insurance. Clearly, regulators must ensure that insurers meet the terms of their contracts. This becomes difficult, however, when these terms are unclear or claimants seek to retroactively expand coverage beyond that stated in their contract or paid for through premiums. For example, there is the issue of whether insurers should pay the cost of rebuilding a structure to meet more stringent building code standards than those in effect when the structure was originally built. Some regulators believe that code enhancement should be covered under standard insurance contracts, but many insurers disagree. Although a retroactive effort to expand coverage may be intended to help claimants, it can subject insurers to unanticipated losses and encourage fraud and excessive risk-taking. In addressing claim disputes, regulators need to be careful to maintain the integrity of policy provisions; failing to do so could further discourage entry, lessen availability, and increase rates.

Regulators have a substantial role to play in facilitating the claims adjustment process following a natural disaster. Regulators in California, Florida, South Carolina, and Texas, among other states, have developed comprehensive disaster response programs that are key to recovery efforts. The NAIC has revised and expanded its disaster response manual based on the experiences of these states. This manual outlines a number of steps that regulators can take to expedite payment of claims from a natural disaster, including providing an information clearinghouse for consumers and insurers, and helping to support adjusters rushed in to deal with the flood of claims. Some states have even taken steps such as controlling the prices of building materials following a disaster.

Other Market Practices

Other market practices that may affect the availability and cost of disaster insurance include sales and marketing activities, appointment and termination of agents, and so on. It is not clear that the regulation of these practices can significantly improve market conditions for disaster insurance. For example, it is very difficult to force insurers to sell coverage to populations in high-risk areas. It also is hard to measure the intensity of insurers' efforts to market to particular areas, and their success will be affected by a number of factors outside their control.

Residual market mechanisms and other state-sponsored disaster risk financing mechanisms (discussed in Chapters 2, 4 and 5) raise some interesting policy questions. Regulators generally believe that these mechanisms can be effective in providing an intermediate layer of excess reinsurance coverage. Residual market mechanisms could promote greater economic efficiency if they facilitate more effective risk pooling within a state, or if they can utilize state government guarantees and recoupment mechanisms to facilitate long-term risk financing, which in turn facilitates the diversification of risk over time.

State residual market mechanisms have been affected by the decreased supply of residential property coverage because of catastrophe risk. Most states have seen small or moderate increases in their residential property RMMs because of catastrophe risk and insurance availability problems, as reflected in Figure 8-4. However, Florida and Hawaii have very large residual markets (representing approximately 30 percent of the market) because of their high catastrophe risk exposure. RMMs of this size are difficult to sustain for a long period, and there are vigorous efforts to depopulate them.

Public insurers may be able to secure favorable tax treatment for catastrophe reserves that may be denied to private insurers because of concerns that private insurers would manipulate reserves to lower their taxes. If state risk pools supplement (rather than supplant) private reinsurance market capacity, they could increase the supply of voluntary market coverage. Such mechanisms cannot be used to enforce cross-subsidies within a state unless they are established as a monopoly, and all property owners are forced to buy insurance. If coverage were voluntary, private insurers would avoid selling, and property owners would be disinclined to buy insurance that was overpriced in order to support subsidies for high-risk areas.

State catastrophe pools have limited capacity to solve disaster insurance availability problems in high-risk states. A state mechanism can only pool exposures within a state; this insufficiently diversifies the risk of very large catastrophes that could occur within a state. States can accumulate tax-favored reserves over time, as well as borrow money if necessary, but the size of their markets still limits the amount of funds they can accumulate or borrow. As noted above, rough estimates of the total catastrophe capacity (including state pools) for Florida or California are $10–$12 billion, which falls considerably short of the potential $50–$75 billion in insured losses that could occur from a catastrophe in a major population center of either state.

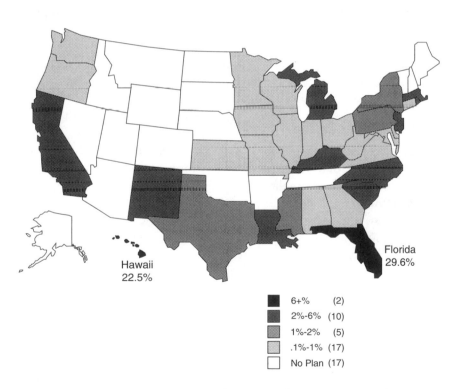

Florida
29.6%

Hawaii
22.5%

■	6+%	(2)
▨	2%-6%	(10)
▦	1%-2%	(5)
▢	.1%-1%	(17)
□	No Plan	(17)

FIGURE 8-4 Residual market share, by state.   Shows percentage of insurance market in each state that is covered by residual market mechanisms.

COORDINATING GOVERNMENT POLICIES
TOWARD DISASTER RISK

Policymakers have an array of options in structuring the government's role in financing disaster risk. It is important to coordinate government actions at various levels to achieve public objectives. Insurance regulators, in particular, must consider how other government actions (e.g., hazard mitigation) affect the behavior of firms and individuals in responding to disaster risk.

Government policy can have significant implications for insurance regulation regarding mandates on the purchase of disaster insurance. On the one hand, such mandates can help to combat adverse selection and expand the pool of properties that are insured, which contributes to risk diversification and the capacity of the system to accommodate large catastrophic events. On the other hand, coverage mandates can significantly increase political pressure on insurers and regulators to make coverage available and suppress prices, and can interfere with market forces. It may be preferable to encourage the purchase of insurance, without mandating it, and to use voluntary mechanisms to ensure the maximum availability and purchase of coverage at the lowest possible price. Of course, this strategy could require the government to be less generous in providing post-disaster aid, which would be politically difficult.

Ultimately, we may be faced with a choice between three different policy sets. One approach would rely heavily on government decisions about determining where people build structures, how they build, how much insurance they are required to buy to pre-fund their own disaster losses, and how much public assistance they will receive following a disaster. However, it is not clear that government management of catastrophe risk is either necessary or would be the most efficient approach. Public choice mechanisms suffer from significant imperfections and are warranted only when there are severe market failures that preclude reliance on private choice and markets.

An alternative policy set would rely entirely on private choice mechanisms to encourage decision makers to choose optimal levels of mitigation and risk diversification through insurance. With this approach, the government would refuse to provide post-disaster assistance to individuals and firms that fail to take adequate pre-disaster precautions. The government also would minimize other kinds of subsidies that discourage individual responsibility in mitigating and financing disaster losses. However, total reliance on private choice might result in suboptimal levels of mitigation and insurance if residents in hazard-prone areas under-estimate the risks facing them and/or do not believe that mitigation and insurance yield public benefits. Some also argue that private markets, at the present time, cannot adequately fund and diversify catastrophe risk by themselves.

A third approach would involve some blend of private and public mechanisms to manage catastrophe risk. The design of a blended model would be tricky, however, because of the conflicts that can occur between those policies that promote *private* choices and those policies that reflect *public* choices about how catastrophe risk should be managed. How to align the incentives for creating a program that combines market mechanisms and government policies is a challenging area for future research.

It would be essential that private incentives for risk mitigation and the purchase of insurance be utilized to the fullest extent possible, using government mechanisms to insure risks that could not be efficiently assumed by the private sector. This implies that the price of insurance, whether provided by private insurers or the government, would be actuarially determined. Government intervention should be restricted to where it is necessary to correct market failures and to where such intervention can effectively promote the public interest. Restrictions on private choices should only be imposed when individuals and firms fail to act prudently. If there is a social preference to subsidize certain groups, such subsidies should be means tested and funded through the most efficient mechanisms available, such as taxes rather than insurance pricing structures.

SUMMARY

Society and consumers rely on the government to oversee insurance markets to protect their interest in obtaining adequate protection against risk at a reasonable price. Insurance regulation is the principal responsibility of the states, although the federal government retains authority to intervene, as it has done on several occasions. The state and federal governments are under considerable pressure to maintain the availability of catastrophe insurance at a price that homeowners can afford. At the same time, insurers must increase their prices for catastrophe insurance and reduce their exposure to risk if they are to remain profitable and solvent in the event of a major catastrophe. This conflict has created significant tension between regulators and insurers in high-risk states and has led to calls for the establishment of supplemental federal catastrophe insurance or reinsurance mechanisms. The current state of affairs is untenable, and there is a need to develop efficient and equitable approaches to managing catastrophe risk, supported by appropriate government policies.

Understandably, regulators and legislators in high-risk states have responded to the current crisis by restricting insurers' rate increases and their attempts to reduce their exposure to catastrophe losses. This policy yields short-run political gains from maintaining the availability of affordable catastrophe insurance, but it does so at the expense of creating market distortions that will exacerbate the catastrophe risk problem in the long run.

Policymakers and regulators face difficult choices, but they must work toward creating a framework that will support reasonable management of catastrophe risk. Different policy sets might be considered, but the one that is most likely to achieve political support would rely on

SIDEBAR 8-1

Questions for Further Research

This discussion of insurance regulation and catastrophe risk raises many more questions than it resolves. Some of the fundamental policy questions that researchers and decision makers need to consider in crafting appropriate regulatory policy are summarized below.

What is the nature of the market problems and market failures with respect to catastrophe insurance that can be addressed by regulation? How closely can regulators monitor and control insurers' catastrophe risk? How should regulators balance the supply of catastrophe insurance with insurers' insolvency risk? To what extent can and should regulators limit insurers' actions with respect to pricing, underwriting, policy design, and other functions for catastrophe insurance? Are regulatory restrictions needed, and can they be enforced without causing market dislocations which retard the market's movement toward an efficient equilibrium? Are there other regulatory approaches that could be considered that would facilitate the effective operation of market forces while ensuring consumer protection where it is really needed? What would happen if regulators removed market constraints in high-risk areas? What are the appropriate role and structure of residual market mechanisms and government-sponsored insurance/reinsurance pools for catastrophe risk and insolvency guaranty funds? These questions will need to be addressed in forging sensible governmental policies with respect to managing catastrophe risk.

private markets to the fullest extent possible, supplemented by government intervention where it is necessary and where it can efficiently correct market failures. The mix of private and public risk management mechanisms could evolve over time, as private markets become able to expand their role in providing catastrophe insurance. There are a number of measures that regulators might consider taking to expand the supply of catastrophe insurance and lower insurers' financial risk. These measures include:

- relaxing entry to and exit barriers from catastrophe insurance markets;
- restructuring solvency regulation to facilitate more efficient assumption and diversification of catastrophe risk;
- allowing insurers to modify policy forms to expand the options available to consumers to share a greater proportion of catastrophe risk if it is efficient for them to do so;
- easing regulation of insurance prices and gradually eliminating attempts to enforce cross-subsidies between low- and high-risk insureds;
- eliminating unnecessary restrictions on insurers' underwriting selection to counter adverse selection and reduce insurers' disincentives to enter catastrophe insurance markets;
- structuring residual market mechanisms, reinsurance funds, and guaranty funds to minimize cross-subsidies and moral hazard and to ensure adequate funding of potential claims obligations; and
- coordinating federal, state, and local government policies toward risk mitigation and disaster assistance in order to maximize private incentives to mitigate risk and purchase adequate insurance.

Legislators and insurance regulators will be reluctant to take such politically difficult steps unless they believe that coordinated private and public sector actions will ensure an adequate supply of insurance coverage at prices no higher than necessary to cover the risk. This may require a federal government commitment to supplying residual catastrophe reinsurance to the extent necessary to secure an adequate supply of primary insurance coverage. Political considerations might require a phased approach to relaxing market constraints so that prices for high-risk areas are gradually increased to levels the market can sustain. Tax-funded subsidies to low-income homeowners in high-risk areas also may be necessary. Any transition strategy should move deliberately toward the ultimate goal of efficient and equitable management of catastrophe risk.

CHAPTER NINE

A Program for Reducing Disaster Losses Through Insurance

HOWARD KUNREUTHER

T HIS CHAPTER FIRST LOOKS at the current market for insurance against natural disasters by addressing the following questions: What are the factors influencing property owners' demand for insurance? Why are insurers reluctant to provide coverage against hurricanes, floods, and earthquakes today? An understanding of both the demand and supply sides of the market suggests a disaster management program where private insurance plays a central role in bringing together other policy tools and key interested parties.

THE CURRENT SCENE

It is apparent from the preceding chapters that residents of hazard-prone areas as well as the insurance community are reluctant to deal with natural disasters today for very different reasons. Many homeowners at risk are not anxious to purchase insurance voluntarily because they feel the disaster will not happen to them; others who have compared premiums with potential benefits may feel that insurance is *not* a good investment, as indicated by the number of policies that have been canceled since the California Earthquake Authority was instituted in California in 1996.

Private insurers have been reluctant to promote coverage against hurricanes, floods, and earthquakes because of uncertainty regarding the risk and because of concern with what the financial consequences of a natural disaster would be for their companies. Hurricane Andrew and the Northridge earthquake were wake-up calls for many firms, alerting them to the possibility of insolvency following another major event. Hence, insurers want to limit their exposure in hazard-prone areas. In some cases they are restricted from doing so by state regulatory agencies and insurance commissioners.

Demand for Insurance

As shown in Chapter 3, many property owners at risk do not purchase insurance voluntarily. As of July 1997 only 20 percent of residential properties in flood-prone areas in the United States were covered by insurance. In California, which is the area in the continental United States most susceptible to earthquakes, less than 20 percent of residences have earthquake insurance, down from 30 percent to 40 percent at the time of the Northridge earthquake. Most homes have protection against wind damage from hurricanes, since it is part of the standard homeowners' policy, which is normally required by financial institutions as a condition for a mortgage.

Unless individuals have personally experienced an event or know others who have suffered losses, they are unlikely to *voluntarily* purchase insurance coverage against natural disasters (Kunreuther et al., 1978). Relatively little insurance was purchased by residents of Kobe prior to the January 1995 earthquake, presumably because the city had never experienced a severe earthquake. Following the severe earthquakes in California in the 1980s and early 1990s, voluntary demand for insurance increased because residents in affected areas became concerned about future damage from these disasters (Palm, 1995).

Figure 9-1 presents a model of the decision-making process involved when individuals choose whether to purchase hazard insurance. Residents in hazard-prone areas have a perceived probability of a disaster p which influences their decision on whether to purchase coverage. There is likely to be a threshold probability p^* below which a person will *not* want to buy coverage because the costs—the attention costs of thinking about the event as well as the transaction costs of collecting information

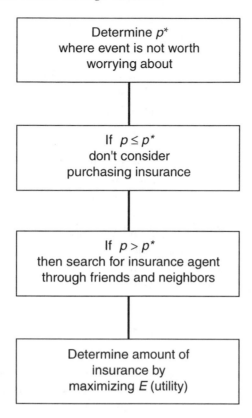

FIGURE 9-1   Sequential model of choice for homeowners' decisions.   Source: Kunreuther et al., 1978.

and deciding on the amount of coverage—are too high. The higher these costs, the larger p^* will be. Those who feel that the probability of a disaster exceeds their cost threshold ($p > p^*$) are likely to seek out friends and neighbors for information on where to buy coverage (Kunreuther et al., 1978; Weinstein, 1987). Once they obtain this information, people base their actual purchase decision on tradeoffs between the costs of insurance in relation to the risks of being unprotected, using a set of simplified rules rather than undertaking detailed calculations.

   In her recent studies of insurance purchase in California, Palm (1995) found that the decision to purchase coverage is influenced primarily by assessing risks and costs. This suggests that most people in California are already aware of the presence of earthquake insurance and do not have to turn to friends and neighbors for information on coverage. It thus appears that the demand for insurance depends on whether the product is at an early or late stage of the diffusion process. In the case of earthquake coverage, the market is much more mature than it was in the 1970s when earlier demand studies were undertaken.

What determines the demand for insurance when individuals have the freedom to specify coverage limits is still not well understood, although recent controlled experimental studies provide insight into consumer decision processes. For example, there is evidence that framing manipulations influence how much consumers are willing to pay for coverage. Some of these factors are the vividness of the media's reporting of a projected event, the use of rebates so that policyholders feel they have experienced a gain if they do not collect on their policies, and the use of the status quo as a reference point for deciding between options. Data from insurance markets indicate that these same effects occur when actual insurance decisions are made (Johnson et al., 1993).

Supply of Insurance

Turning to the supply side of the market, Chapter 2 points out that insurers and reinsurers are concerned about the impact of future severe earthquakes and hurricanes on their solvency. Large insured losses from such disasters could significantly reduce their surplus, which is likely to lead to a restriction in future coverage. Firms in the industry are therefore considering either rationing policies written in high-risk areas and/ or supplementing traditional reinsurance with funding from the capital markets.

The literature in recent years suggests that insurers and other firms are risk averse, and hence they are concerned with non-diversifiable risks such as catastrophic losses from disasters (Mayers and Smith, 1982). (See Froot et al., 1994, for a discussion of the increasing role that risk management tools are playing in how corporations are doing business today.) As discussed in Chapter 2, insurers are also likely to be averse to ambiguity, as shown by their taking uncertainty regarding loss into account when determining the premium they would like to charge. Actuaries and underwriters both utilize heuristics that reflect these concerns.

Actuaries normally determine a premium based on expected value by assuming that the probability and loss are known. They then increase this value to reflect the amount of perceived ambiguity in the probability and/or uncertainty in the loss. One commonly used formula for determining a premium is $z = (1 + \lambda)\mu$, where μ = expected loss (i.e., $p \times L$) and $\lambda > 0$ is a factor reflecting ambiguity and uncertainty independent of any adjustment to cover administrative costs (Lemaire, 1986).

Underwriters decide whether a risk is insurable by utilizing the actuary's recommended premium z as a reference point, and then focus on the impact of a major disaster on their company's solvency. In other words, underwriters are first concerned with the firm's safety and then with profit maximization. A safety-first model, first proposed by Roy in 1952, can be contrasted with a value maximization approach to firm behavior. A safety-first model explicitly concerns itself with insolvency when making a decision regarding maximum amount of coverage and premiums to charge. A value maximization model recognizes that firms are risk averse so that premiums have to be higher to reflect the chances of a catastrophic loss. It does *not* explicitly focus on keeping the probability of insolvency below some prespecified level. Stone (1973) formalized these concepts by suggesting that an underwriter who wants to determine the conditions under which a specific risk is insurable will first focus on keeping the probability of insolvency below some threshold level q^*.

As an example, suppose that an insurer expects to sell S policies, each of which can create a loss L. The underwriter recommends a premium z^* so that the probability of insolvency is no greater than q^*. Risks with more uncertain losses or greater ambiguity will cause underwriters to want to charge higher premiums for a given portfolio of risks. The situation will be most pronounced for highly correlated losses, such as earthquake policies sold in one region of California.

A safety-first model of underwriter behavior is consistent with the Mayers and Smith (1990) rationale as to why insurance firms want to purchase reinsurance. In fact, a rule that focuses on keeping the chances of insolvency below q^* explicitly recognizes the role that risk plays in the decision process. By characterizing the underwriter's behavior in this way, one can then examine how reinsurance and other capital market instruments can alleviate these concerns.

Interview data with several U.S. insurance companies provides additional evidence that firms follow a safety-first model. Prior to Hurricane Andrew (1992) and the Northridge earthquake (1994), these insurers were not worried about the potential impact of losses to their portfolio from severe hurricanes and earthquakes. Hence they did not attempt to restrict coverage and/or make the case for higher premiums because of the likelihood that they would become insolvent.

In the aftermath of Hurricane Andrew and Northridge, company executives modified their views and are now concerned about their ability to survive a future catastrophe given their current portfolios and the amount of reinsurance coverage they can obtain at a reasonable price. In other words, they feel that their chance of insolvency based on their current portfolios exceeds their threshold level of concern q^*. Therefore, they want to exercise one or more of the following options: reduce the number of policies they write in catastrophe-prone areas, raise their per-unit premium z, or obtain more reinsurance coverage.[1] Earlier chapters in this book have shown that insurers in the United States have had problems undertaking these actions due to regulatory constraints in some states (e.g., Florida), which have set ceilings on the premiums they can charge and limited the number of policies that they cancel in any given year.

Winter (1988, 1991) and Doherty and Posey (1992) have shown that a particularly severe earthquake or hurricane could have a very negative impact on the availability of insurance throughout the country. Doherty, Posey, and Kleffner (1992) have examined how insurers have responded to a variety of surplus shocks in the past. Their analysis suggests that only 50 percent of the lost surplus is likely to be replaced following a catastrophic loss, so that the availability of coverage in many different lines of insurance would have to be reduced in the immediate aftermath of the disaster.

Key Role of Insurance

The challenge society faces today is how to promote investments in cost-effective risk reduction measures, while at the same time placing the burden of recovery on those who suffer losses from natural disasters. *In theory,* insurance is one of the most effective policy tools for achieving both objectives because it rewards investments in cost-effective mitigation with lower premiums and provides indemnification should a disaster occur. (This presumes that both homeowners and insurers are aware of state-of-the-art technologies and can determine what impact they will have on reducing expected losses from future disasters.)

---

[1]These observations are based on a series of personal interviews with insurers and reinsurers conducted by Jacqueline Meszaros as part of a National Science Foundation study, "The Role of Insurance and Regulation in Dealing with Catastrophic Risk" (1997).

In practice, insurance has not played this role in recent years. Insurers generally do not charge premiums that encourage loss prevention measures for several reasons. First, they feel that few people would voluntarily adopt these measures based on the small *annual* premium reductions in relation to the large up-front cost of investing in these measures. If individuals have short time horizons, they would have little interest in investing $1,200 in return for a reduction in premiums of $200.

As shown in Chapter 8, insurance is a highly regulated industry; rate changes and new policies generally require the approval of state insurance commissioners. Premium schedules with rate reductions for adopting certain mitigation measures require administrative time and energy both to develop and to promote to the state insurance commissioners. If potential policyholders do not view mitigation measures as attractive investments, insurers who developed these premium reduction programs would be at a competitive disadvantage relative to those firms who did not.

A HAZARD MANAGEMENT PROGRAM FOR DEALING WITH CATASTROPHE RISKS

Private insurance can be an important part of a hazards management program, but this requires a reorientation of its role in preventing losses as well as covering damage from disasters. Fortunately, we can think more broadly about how insurance can help manage these risks in the future because of advances in the technology for analyzing data, and because of the recent availability of capital market funding for supplementing traditional reinsurance. These developments suggest a strategy in which private insurance plays a key role in encouraging cost-effective risk reduction measures while providing protection against the financial consequences of disaster.

For such a strategy to be successfully implemented, the insurance and reinsurance industries need to be closely linked with other interested parties in the private sector, notably the financial institutions, investment bankers, and the building and real estate communities. Also, government agencies at the local, state, and federal levels need a much clearer understanding of the importance of having insurance rates reflect the risks of living in hazard-prone areas, and of the possible need for multistate and/or federal arrangements to cover some of the losses from future disasters. The proposed hazard management program consists of the following elements, which will be discussed in more detail below:

- improving estimates of risk
- auditing and inspecting property
- emphasizing building codes
- providing economic incentives for mitigation
- broadening protection against catastrophic losses.

Improving Estimates of Risk

Insurers will benefit from improved estimates of the risk associated with catastrophes in two ways. First, by obtaining better data on the probabilities and consequences of disasters, insurers will be able to more accurately set their premiums and tailor their portfolios to reduce the chances of insolvency. The improved information should enable them to more accurately determine their needs for protection through reinsurance or capital market instruments. Second, more accurate data on risk also reduces the asymmetry of information between insurers and other providers of capital. Investors are more likely to supply additional capital as they become increasingly confident in the estimates of the risks of insured losses from natural disasters.

In setting rates for catastrophe risks, insurers have traditionally looked backwards, relying on historical data to estimate future risks.[2] This process is likely to work well if there is a large database of past experience from which to extrapolate into the future. Low-probability, high-consequence events such as disasters, by their very nature, make for small historical databases. Thus, there is a need to integrate scientific estimates of the probabilities and consequences of events of different magnitudes with the evidence from past experience. (New advances in seismology and earthquake engineering are discussed in FEMA, 1994, and Office of Technology Assessment, 1995.)

In many cases some controversy exists among the experts as to the risks from hazards and technologies. For example, scientific knowledge about the probability of earthquakes of different magnitudes has been growing rapidly since the 1960s, but there is still no consensus on what information to use as the basis for seismic probability maps (Mittler et al., 1995).

Advances in information technology have encouraged catastrophe modeling, which can simulate a wide variety of different scenarios reflecting the uncertainties in different estimates of risk. For example, it is now feasible for insurers to evaluate the impact of different exposure

---

[2]I am grateful to Terry van Gilder of Risk Management Solutions, formerly chief underwriter at Chubb, for characterizing the decision process of insurers in this way.

levels on both expected losses and maximum possible losses by simulating a wide range of different estimates of seismic events using the data generated by scientific experts. Similar studies can be undertaken to evaluate the benefits and costs of different building codes and loss prevention techniques (Insurance Services Office, 1996).

The growing number of catastrophe models has presented challenges to users who are interested in estimating the potential damage to their portfolios of risks. Each model uses different assumptions, different methodologies, different data, and different parameters in generating projections. The models' conflicting results make it difficult for the insurer to know what premiums to set to cover their risks; the results also make it difficult for reinsurance and capital market communities to feel comfortable investing their money in providing protection against catastrophe risk.

Hence the need for a better understanding of why these models differ and of the importance of reconciling these differences in a more scientific manner than has been done up until now. Bringing the leading modelers together with the insurers, reinsurers, and capital markets to discuss how their data are generated may reduce the mystery that currently surrounds these efforts.

Auditing and Inspecting Property

One way to determine the ability of a structure to withstand the impact of natural disasters would be to inspect the property carefully. A

BOX 9.1 Open Question about Estimating Risk

- Are hazard- and loss-estimating modeling approaches sufficiently reliable and valid measures to guide the underwriting decision process?
- How can one better characterize the uncertainties in determining the probability of disasters of different magnitudes and the vulnerability of structures to these events?
- How can one improve the database for assessing risks in different geographical areas and reduce uncertainty about the vulnerability of different types of structures to the natural hazards to which they are exposed?
- What information does the insurance industry want to better assess risks, and how does the industry anticipate obtaining this information in the near future?
- How will the additional mapping efforts (e.g., seismic mapping in California and Oregon) aid insurers in developing rate schedules for natural hazard risks?

careful appraisal of the structure is expensive. To date, such audits have been undertaken primarily on commercial risks where the insurer has absorbed the cost of the audit through its large premium base with the policyholder. With respect to residential properties at risk, one way to encourage the adoption of cost-effective risk reduction measures would be to incorporate them in building codes and provide a seal of approval to each structure that meets or exceeds these standards. (Kunreuther and Kleffner, 1992, provide a rationale for strengthening building codes by analyzing the factors that lead individuals to avoid investing in mitigation measures even if they have accurate information on the risk.)

Banks and financial institutions could require that structures be inspected and certified against natural hazards as a condition for obtaining a mortgage. This inspection, which would be a form of buyer protection, is similar in concept to termite and radon inspections normally required when property is financed. Natural hazard inspections are not routinely undertaken by banks today, even though it is in their interest to know as much about the risk as possible to protect their mortgages. The success of a program of natural hazard inspection requires the support of the building industry, of realtors, and of a cadre of inspectors, well qualified to provide accurate information on the condition of the structure.

Emphasizing Building Codes

Building codes serve two principal purposes: they help correct misimpressions that property owners have with respect to the safety of their structures, and they reduce disaster losses that are directly attributable to the collapse of buildings.

How Building Codes Promote Accurate Information about Risk

One reason why property owners misperceive risks is that the real estate community has limited interest in providing information on the nature of hazards, even when required to do so by legislation such as the Alquist-Priolo Special Studies Zone Act (Palm, 1981). Furthermore, engineers and builders have limited economic incentives for designing safer structures since doing so normally means incurring costs that they feel will hurt them competitively (May and Stark, 1992).

Well-enforced building codes help correct misinformation that property owners may have regarding structural safety, while leveling the playing field for constructing buildings. For example, suppose a property

---

BOX 9-2   Open Questions about Property Inspection

- How accurate are inspections of existing buildings in determining whether risk reduction measures are in place?
- Can the developer be made responsible for installing risk reduction measures on new buildings by inspecting the structure at different stages of its construction?
- How could one design a seal of approval that is understandable to the potential home buyer?

---

owner believes that hurricane damage to a structure would total $20,000, but the developer knows that the total would be $25,000 because the house is not well constructed. The developer has no incentive to relay the correct information to the prospective property owner because the developer is *not* held liable for hurricane damage to a poorly constructed building. If the insurer is unaware of how well the building is constructed, the insurer cannot convey this information to the potential owner through risk-based premiums. Inspecting the building to see that it meets code and then giving it a seal of approval provides accurate information to the property owner. It also forces the developer and real estate agent to let the potential buyer know why the structure has not been officially approved for safety.

Officially designating structures in this way has an additional side benefit. Insurers may want to limit coverage only to those pieces of property that receive a seal of approval. Evidence from a July 1994 telephone survey of 1,241 residents in six hurricane-prone areas on the Atlantic and Gulf Coasts provides evidence supporting this type of program. Over 90 percent of the respondents felt that local home builders should be required to follow building codes, and 85 percent considered it very important that local building departments conduct inspection of new residential construction (Insurance Institute for Property Loss Reduction, 1995).

The Insurance Institute for Property Loss Reduction (IIPLR), now the Institute for Business and Home Safety, has been a driving force in providing economic incentives for communities to adopt building codes and in advocating code enforcement. Formed by the property-casualty insurance companies in the United States, this independent, nonprofit organization helped create the Building Code Effectiveness Grading Schedule (BCEGS). This rating system, administered by the Insurance Services Office, measures how well building codes are enforced in communities around the United States. Property located in communities that have well-enforced codes will presumably benefit through lower insurance premiums. For fire insurance, lower premiums are based on the level of fire protection.

How Building Codes Reduce Externalities

Cohen and Noll (1981) provide an additional rationale for building codes. A building collapse may create externalities in the form of economic dislocations and other social costs in addition to the economic loss suffered by the owners. The owners may not have taken these consequences into account when evaluating specific mitigation measures.

Consider the following example. A building toppling off its foundation after an earthquake could break a pipeline and cause a major fire, which could damage other homes that had not been affected by the earthquake in the first place. This type of damage has a direct impact on the pricing of insurance in a hazard-prone area. Suppose that an unbraced structure that toppled in a severe earthquake had a 20 percent chance of bursting a pipeline and creating a fire which would severely damage 10 other homes, each of which would suffer $40,000 in damage. Had the house been bolted to its foundation, this series of events would not have occurred.

The insurer who provided coverage against these fire-damaged homes under a standard homeowner's policy would then have had an additional expected loss of $80,000 (i.e., $.2 \times 10 \times \$40,000$). A building code requiring concrete block structures to be braced in earthquake-prone areas would have prevented this loss. One option for dealing with this situation would be for homes adjacent to those that are not braced to be charged a higher fire premium to reflect the additional hazard of living next to the unprotected house. Charging this additional premium to the unprotected structure that caused the damage would be more equitable, but this cannot legally be done. Thus, each of the 10 homes that are vulnerable to fire damage from the quake would be charged this extra premium.

The point of this analysis is that mitigation yields an additional annual expected benefit over and above the reduction in losses to the specific structure adopting a mitigation measure. All financial institutions

---

BOX 9-3   Open Questions about Building Codes

- What empirical studies are necessary for determining the magnitude of the social costs and externalities that could be reduced through well-enforced building codes?
- Can we utilize past experience and engineering studies on building performance to evaluate the cost-effectiveness of mitigation measures with sufficient precision that they can be incorporated into building codes?
- Will the use and enforcement of well-specified codes be sufficient to change the underwriting practices of the insurance industry?
- What types of penalty systems for structures which do not meet code would impel developers to adhere to building code standards when building new construction?

---

and insurers who are responsible for these other properties at risk would favor building codes to protect their investments and/or reduce the insurance premiums they charge for fire following earthquake.

Providing Economic Incentives for Mitigation

As pointed out in Chapter 7, insurers could provide financial incentives in the form of lower premiums, lower deductibles, or increased coinsurance to encourage the adoption of cost-effective risk reduction measures. If budget constraints prevent a property owner from investing in these mitigation measures, then insurers should consider having a bank or other financial institution provide funds through a home improvement loan with a payback period coterminous with the life of the mortgage.

Consider the following example, where the cost of risk reduction measures on a piece of property in earthquake-prone country is $1,500. If the seismologists' best estimate of the annual probability of an earthquake is 1/100, and the reduction in loss from investing in the risk reduction measure is $27,500, then the expected annual benefit is $275. A 20-year loan for $1,500 at an annual interest rate of 10 percent would result in payments of $145 per year. If the annual insurance premium reduction reflected the expected benefits of the risk reduction measure (i.e., $275), then the insured homeowner will have lower *total* payments by investing in mitigation than by not undertaking the measure.

The above example also illustrates the robustness of the risk estimates that would still make it desirable for a homeowner with insurance to take out a long-term loan for mitigation payments. Even if the annual probability of an earthquake is as low as 1/189, the property owner would want to take out the loan. (If the probability of an earthquake were 1/189, the annual premium reduction would be $145, the same as the annual loan payment.) Similarly, if the probability p were known to be 1/100, then the reduction in loss from mitigating could be as low as $14,500, and the loan would still be attractive because of the benefits from the reduced insurance premium.

Many poorly constructed homes are owned by low-income families who cannot afford the costs of mitigation measures or the costs of reconstruction should their house suffer damage from a natural disaster. Equity considerations argue for providing this group with low-interest loans and grants so that they can either adopt cost-effective risk reduction measures, or relocate to a safer area. Since low-income victims are likely to receive federal assistance to cover uninsured losses after a disaster, subsidizing these mitigation measures can also be justified on efficiency grounds.

BOX 9-4 Open Questions on Economic Incentives for Mitigation

- Is there sufficient knowledge about natural hazard risks to incorporate the effectiveness of mitigation alternatives into the insurance underwriting process?
- Do insurance companies have the ability to assess the quality of retrofit as a part of the insurance underwriting process to determine what premium reductions these actions merit?
- Will insurers utilize the community rating programs in their underwriting processes for natural hazard insurance?
- What types of economic incentives aside from premium reductions (e.g., lower deductibles, higher limits of coverage) are likely to be attractive enough to policyholders to encourage them to adopt mitigation measures?
- What would be the most effective ways of providing subsidies to low-income families to encourage them to adopt cost-effective risk reduction measures?

Broadening Protection Against Catastrophic Losses

New sources of capital from the private and public sectors could provide insurers with funds against losses from catastrophic events, which would alleviate insurers' concerns that the next major disaster might leave them insolvent. This section examines some of these new developments and proposals.

Role of the Capital Markets

In the past couple of years investment banks and brokerage firms have shown considerable interest in developing new financial instruments for protecting against catastrophe risks (Jaffee and Russell, 1996). Their objective is to find ways to make investors comfortable trading new securitized instruments covering catastrophe exposures, just like the securities of any other asset class. In other words, catastrophe exposures would be treated as a new asset class.

Some of the recently proposed instruments provide funds to the insurer should it suffer a catastrophic loss. J. P. Morgan and Nationwide Insurance successfully negotiated such a transaction, whereby Nationwide borrowed $400 million from J. P. Morgan and placed the money in a trust fund composed of U.S. Treasury securities. Nationwide pays a higher-than-normal interest rate on these funds in return for having the ability to issue up to $400 million in surplus notes to help pay for the losses should a catastrophe occur (Mooney, 1995).

In June 1997 the insurance company USAA floated Act of God bonds that provided it with protection should a major hurricane hit Florida. In July 1997, a two-year catastrophe bond based on California industry losses was put together by Swiss Re Capital Markets, and Credit Suisse First Boston to deal with catastrophic earthquake losses. This multiyear catastrophe bond compares favorably with the rate of return on high-yield corporate bonds. Other financial arrangements, such as *catastrophe insurance futures contracts* and *call spreads* introduced by the Chicago Board of Trade (CBOT) in 1992, enable an insurer to hedge against its underwriting risk by attracting capital from insurance and non-insurance segments of the economy (Cummins and Geman, 1995; Harrington et al., 1995). The Catastrophic Risk Exchange (CATEX) creates a marketplace where insurers, brokers, and the self-insured can swap units of their catastrophe risks by region and peril. For example, an insurer could swap units of California earthquake for Florida windstorm (Insurance Services Office, 1996).

Use of Insurance Pools

The National Conference of Insurance Legislators (NCOIL) has considered a multistate Natural Disaster Compact, modeled after the Florida Hurricane Catastrophe Fund, to expand the pool, increase available resources, and further spread geographic risks. This concept has obvious appeal to the most disaster-prone states, and an equal lack of appeal to states where disasters are more rare. As pointed out in Chapter 8, these pools face a number of legal and political challenges which may make it difficult for them to be initiated.

A successful example of the use of an insurance pool is the one that provides coverage against catastrophic losses from nuclear power plant accidents in the United States. Under the Price-Anderson Act, a group of private insurers agreed to provide coverage to utility companies for losses that can total up to $8.5 billion (P.L. 100-408) (U.S. Congress, 1995). The German Pharmaceutical Pool, consisting of private insurers and reinsurers from all over Europe, operates in a similar fashion, providing protection against large risks by private drug manufacturers associated with new drugs (Kunreuther, 1989).

Proposed Federal Solutions

Lewis and Murdock (1996) developed a proposal that the federal government offer *catastrophe reinsurance contracts*, which would be auctioned annually. The Treasury would auction a limited number of excess-of-loss (XOL) contracts covering industry losses between $25 billion and $50 billion from a single natural disaster. Insurers, reinsurers, and state and national reinsurance pools would be eligible purchasers. XOL contracts would be sold to the highest bidder above a base reserve price which is risk-based. Half of the proceeds above the reserve price would go into a mitigation fund, with the remainder retained to cover payouts. This federal reinsurance effort would be part of a broader program involving mitigation and other loss reduction efforts.

Another proposed option is for the federal government to provide reinsurance protection against catastrophic losses. Private insurers would build up the fund by being assessed premium charges in the same manner that a private reinsurance company would levy a fee for excess-loss coverage or other protection. The advantage of this approach is that resources at the federal government's disposal enable it to cover catastrophic losses without charging insurers the higher-risk premium that

---

BOX 9-5   Open Questions about Catastrophic Loss Protection

- What impact will new capital market instruments have on the pricing and availability of conventional reinsurance?
- What are the concerns of investors, and how can they be addressed so that they provide more funds, enabling new capital market instruments to expand?
- What are the challenges in developing XOL contracts to supplement reinsurance and capital market instruments? How difficult is it to specify the minimum and maximum retention points so that the private market initiatives can thrive?
- Are there lessons from the National Flood Insurance Program that can be helpful in designing similar programs which would manage hurricane and earthquake risks better than we do today?

---

either reinsurers or capital market instruments would require. If one views the private sector as the first line of attack on the problem, as we do, then one would only want to use federal reinsurance as last resort.

SUMMARY AND SUGGESTIONS FOR FUTURE RESEARCH

This concluding chapter has proposed a new approach for managing natural hazards which stresses the importance of private insurance as a catalyst for reducing losses in the future and covering much of the damage after a disaster occurs. This new approach takes advantage of recent developments in information technology and the emergence of new capital market instruments to deal with non-diversifiable catastrophe risks. These two major changes open up opportunities for residents and firms to undertake cost-effective loss protection measures, while at the same time providing a financial cushion to insurers concerned with the possibility of insolvency. Insurance should thus be able to play a more important role in the future in helping to manage catastrophe risks.

The success of the proposed hazard management program requires the active involvement of a number of interested parties from the private sector such as insurers, banks and financial institutions, realtors, builders, and contractors. It also requires that government officials enforce building codes. Public sector agencies have a role in providing assistance

*Fire in the Marina district, San Francisco, triggered by ground shaking
from the Loma Prieta earthquake, 1989, which ruptured underground
water and gas pipelines (USGS, W. Hays).*

to low-income families so that they can adopt cost-effective mitigation
measures, and so that they can recover after a disaster. The federal government may want to issue XOL contracts or provide catastrophe reinsurance to insurers if the private sector does not offer sufficient coverage.
Interested parties in the private sector need to work closely with public
sector agencies at the federal, state, and local levels to determine a set of
programs that satisfy both the efficiency and equity concerns outlined in
Chapter 1.

    With respect to future directions for research, the types of cost-
effective mitigation measures that could be applied to new and existing
structures should be specified. Only then can insurers, builders, and financial institutions work together to incorporate these measures as part
of building codes and provide property owners with appropriate rewards for adopting them. Questions remain about the use and enforcement of building codes, and about the types of incentives insurers can
provide to individuals who invest in loss mitigation measures (see Boxes
9-3 and 9-4).

    One strategy for developing a disaster management program involves
analyzing the impact of disasters or accidents of different magnitudes on

different structures. Long-term simulations could help in estimating expected losses from these events and in projecting the maximum probable losses arising from worst-case scenarios. By constructing simulations of large, medium, and small *representative* insurers with specific balance sheets, types of insurance portfolios, and premium structures, one could examine the impact of different disasters on the insurers' profitability, solvency, and performance under different scenarios. This simulation exercise would enable one to evaluate how risk reduction measures and the provision of funds against catastrophic losses by reinsurers and the capital markets affect insurers' profitability and likelihood of insolvency. An example of the application of such an approach to a model city in California facing an earthquake risk can be found in Kleindorfer and Kunreuther (in press).

Such an analysis may also enable one to compare how attractive capital market instruments are relative to reinsurance for different types of insurers who have specific risks in place. For example, the recent Act of God bonds issued by USAA are similar in form to a proportional reinsurance contract above a retention level. From USAA's point of view, the bonds may be priced more attractively than a comparable reinsurance contract. (Doherty, 1997 provides more details on these and other recent financial instruments.) One would expect the price of reinsurance to fall in the future, given these and other financing and hedging instruments against catastrophe risk, unless certain features of reinsurance would prevent the price from declining. The data from the simulations could also be used to determine the return an investor would require to provide capital for supporting each instrument. The selling prices of different types of capital market instruments would reflect both the expected loss and variance in these loss estimates to capture risk aversion by investors. (See Froot, 1997, for a detailed discussion of reasons why insurers may view reinsurance as more desirable than capital market instruments.)

The type of simulation modeling we describe must rely on solid theoretical foundations in order to delimit the boundaries of what is interesting and implementable in a market economy. Such foundations will also apply to research on the traditional issues of capital markets and the insurance sector, and to research on the processes by which (re-)insurance companies, public officials, and property owners determine levels of mitigation, insurance coverage, and other protective activities. In the area of catastrophe risks, the interaction of these decision processes, which are central to the outcome, seem to be considerably more complicated than in other economic sectors, perhaps because of the uncertainty and ambiguity of the causal mechanisms underlying natural hazards and their mitigation.

Finally, public sector damage from natural disasters results in a large cost to taxpayers. Government officials should be encouraged to purchase insurance for public structures and invest in cost-effective risk reduction measures. One way to do this is to change legislation so that a lower percentage of damage to these structures is covered by federal disaster recovery funds. Recovery funds would not be available unless municipalities implemented cost-effective mitigation measures. Another alternative is to levy property taxes on all community residents to cover losses to public structures from disasters. This is a form of community-based insurance, with all residents paying a share in proportion to the value of their property. To our knowledge, this alternative has not been seriously proposed as part of future legislation. However, if there is a feeling in Congress that the responsibility for disaster recovery lies primarily with the private sector or at the local level, then insurance, incentives, taxes, and well-enforced local building codes will have a higher profile in the future than they do today. These issues related to the public sector deserve serious study in the future.

This is a very exciting time for the insurance and reinsurance industries to explore new opportunities for dealing with catastrophe risks. If insurance can be used as a catalyst to bring other interested parties to the table, it will have served an important purpose in helping both the industry and society deal with the critical issue of reducing losses and providing protection against damage from earthquakes, floods, hurricanes, and other natural disasters.

Commercial Insurance

EUGENE LECOMTE

T HIS APPENDIX DESCRIBES the insuring of commercial property against natural hazards such as earthquakes, windstorms, and floods. It reviews the insurance coverage for commercial structures and business personal property in general, and discusses a variety of interacting factors that influence the insuring of commercial structures and their contents. It describes how this insurance is marketed (i.e., by agents, exclusive agents, and brokers) by examining the roles of the major players, including "risk managers."

Commercial property insurance policies provide indemnification to the policyholder for direct damage to insured structures and business personal property. They can also insure consequential loss such as that caused by the interruption of business. Direct damage to insured structures and business personal property includes payment for the repair or replacement of the damaged property. Business interruption coverage generally provides indemnity for the loss of net income and continuing expenses and helps to assure that the business will survive the repair period.

PROPERTY COVERED

Most types of property owned by a business are, or can be, covered under a commercial property insurance policy. For instance, the following types of property may be covered:

- building structures
- non-building structures
- business personal property:
  — furniture
  — fixtures
  — equipment
  — supplies
  — machines and machinery
  — stock (raw, in-process, and finished).

EXCLUDED PROPERTY

To avoid disputes, there should be a clear understanding of what the policy covers and what it excludes. Most commercial property insurance policies exclude:

- foundations and other underground property (i.e., pipes and drains)
- grading, excavations, and filling
- plants, lawns, trees, shrubs, growing crops, and land
- paved surfaces, roads, bridges, piers, wharves
- detached signs, antennas, fences, and other outdoor items
- building glass
- retaining walls not a part of a building
- vehicles licensed for road use, watercraft, and aircraft.

BASIC CONCEPTS

Commercial property insurance policies also exclude any property specifically insured under another policy, as well as property sold by the insured under conditional sales installment plans, or other deferred billing plans after delivery to the customer. Also excluded are money, securities, precious metals, and bullion, as well as fine arts, jewelry, watches, furs, and silverware. Excluded items are often covered by endorsement. Again, the important point is to know what is covered, and what is not.

Additionally, in evaluating the coverages, consideration must be given to limits of liability (amounts of insurance), deductibles, and coinsurance clauses.

Starting with basic definitions, there is confusion at the outset, because although all buildings are structures, not all structures are buildings. Examples of non-building structures include silos, water towers and tanks, swimming pools, oil tanks, wharves and piers, bridges, covered bridges, and enclosed walkways—all of which may be covered under commercial insurance. Adding to the confusion is the fact that architects, engineers, and underwriters do not define or classify structures similarly. This appendix includes a discussion of building classifications and relates them to underwriting and developing premiums for different structures. The appendix also discusses the occupancy and location of buildings.

Insuring commercial structures and their contents involves the need for specialized knowledge regarding the construction of the structures and the equipment, machinery, and furniture they contain. It also involves the development of tailored or manuscript coverages, the use of coinsurance clauses and replacement cost provisions, as well as the adoption of underwriting techniques, tools, or devices that accommodate the peculiarities of these structures and their occupancies. Additionally, the underwriting process must consider the perils to be insured, amounts of insurance (exposure values) to be provided, and types of coverage.

WHAT'S AT RISK

The total value in 1993 of all commercial structures in the continental United States was $11 trillion, which represents an increase of 190 percent from 1980, and a 65 percent increase over 1988. The 1993 value of residential and commercial property in the first tier of coastal counties from Maine to Texas, a band of real estate approximately 50 miles in width, was $3.15 trillion. This represents an increase of 178 percent from 1980, and 69 percent over 1988. As the population of the United States continues to grow, and as more and more people settle in the Sun Belt and recreational areas, the value of residential and commercial structures will likewise grow.

Property valued at $1.507 trillion, or 14 percent of the total value of all commercial property in the United States, is situated in the Atlantic and Gulf Coast counties. These properties sit directly in the path of potential hurricanes, severe windstorms, and the ravages associated with storm surge and sea level rise. The remaining 86 percent of the 1993 values of commercial structures are generally found in inland metropolitan centers and along the West Coast. Many of these locations are subject to earthquakes, severe windstorms, mud slides, floods, tornadoes, wildfires, and hailstorms. No property is immune to the ravages of Mother Nature.

These figures refer to the value of buildings and other structures and do not include the value of contents—that is, furniture, equipment, stock, merchandise, raw materials for work in process, or finished goods. Once rain, snow, dirt, and dust penetrate a building's roof and exterior walls, significant damage to the contents can follow. Costly building contents, such as computers, electric equipment, and manufacturing machinery, are especially vulnerable to loss caused by exposure to natural hazards. While structural values are significant, they represent only a portion of the values to be insured, and the value of contents will rapidly raise the overall values to be insured. Adding to the financial burdens of commercial insurers are the losses that can occur in connection with business interruption, general liability, workers' compensation, and automobile, glass, burglary, and theft insurance.

Over the years commercial property insurance has been widely available, with the exception of insurance against losses related to asbestosis and environmental clean-up, that is, losses associated with the Superfund program. Competition for commercial business has been keen, and regulatory intervention has been practically nonexistent compared to that for the personal lines (homeowners' and private passenger automobile insurance). Will these situations change? Will the commercial property insurance markets be adversely affected by the following issues:

- The failure of steel frame buildings as evidenced in the Northridge earthquake?
- The vulnerability of structures with External Insulation and Finish System (EIFS) siding to wind and rain during hurricanes?
- The apparent increase in frequency and severity of natural disasters (i.e., hurricanes, tornadoes, floods, and wildfires)?
- A potential for greater rates of precipitation (rain and snow) caused, perhaps, by climate change?

Can mitigation programs and techniques and new loss reduction technologies have a salutary impact on commercial insurance? Can new financial alternatives such as Chicago Board of Trade Futures, Act of God bonds, and state catastrophe pools assist in pricing of insurance policies by providing the required financial cushion against large losses? If not, how can the solvency of insurers be preserved and availability problems avoided?

UNDERWRITING COMMERCIAL PROPERTY

The underwriting of a commercial building involves a meticulous analysis of its construction, physical characteristics, condition, occupancy, and location. In each instance, the inquiry must be thorough if the loss exposure is to be understood and a proper premium established. As discussed in Chapter 2, not all situations are insurable. In deciding questions of insurability, the actuary and underwriter must make both statistical and administrative (business) judgments. Reasonable exceptions may be made if the insurer has imposed practical controls through underwriting, policy provisions, and other means to compensate for what might otherwise be uninsurable.

In the insuring of commercial property, special attention must be paid to the conditions of insurability because of high values, the potential for more frequent and severe events, and the shrinking of the supply of affordable reinsurance.

Classification of Structures

Commercial buildings include the following categories of structures: high-rise, reinforced concrete, heavy steel, reinforced masonry, light steel, timber, and unreinforced masonry.

There is no uniform definition for commercial structures that may be used by the various stakeholders. For insurance underwriting and rating purposes the following definitions set forth in the Insurance Service Office's Commercial Lines Manual (ISO-CLM) are used:

- *Frame:* Exterior walls of wood, combined with brick veneer, stone veneer, wood iron-clad, stucco on wood.
- *Joisted Masonry:* Exterior walls of masonry material (adobe, brick, concrete, gypsum block, hollow concrete block, stone, tile, or similar materials) with combustible floor and roof.
- *Noncombustible:* Exterior walls, floors, and roof constructed of metal, asbestos, gypsum, or other noncombustible materials.
- *Masonry Noncombustible:* Same as joisted masonry except that the floors and roof are of metal or other noncombustible materials.

- *Modified Fire Resistive:* Exterior walls, floors, and roof of constructed masonry or fire-resistive material with a fire resistance rating of at least one hour, but less than two hours.
- *Fire Resistive:* Exterior walls, floors, and roof of masonry or fire-resistive material with a fire resistance rating of at least two hours.

These classifications reflect the emphasis placed on insuring loss caused by fire. They do not address the quality of construction or consider the applicable building codes. Under the circumstances, underwriters must probe other avenues of inquiry to ascertain that engineered structures have met at least the standards prescribed by the local building, fire, and electric codes.

The ISO-CLM provides three classifications for wind and seven for earthquake. The wind categories are as follows:

1. *Wind Resistive:* Modified fire resistive and fire resistive;
2. *Semi–Wind Resistive:* Masonry noncombustible; and
3. *Ordinary:* Frame, joisted masonry, and noncombustible.

These classifications give little or no indication of a structure's ability to withstand the forces of wind. They provide no information about the structure's ability to resist the vertical or lateral wind loads or about the building codes under which the structure was erected.

The seven earthquake building classifications identified in the ISO-CLM are:

1. Wood Frame
2. All Metal
3. Steel Frame
4. Reinforced Concrete, Combined Reinforced Concrete, or Structural Steel
5. Concrete Brick or Block
6. Earthquake Resistive
7. Special Structures.

Each class and its applicable subclasses are described in the ISO-CLM. The adequacy of these classifications for identifying the earthquake vulnerabilities of buildings is questionable. It would seem that to reduce the uncertainties, these classifications should be related to information on soil conditions and distance to the fault line. Has the time arrived when a new set of fire, wind, and earthquake building classifications is needed? Should these new classifications be constructed so that they are compatible with some type of model code, such as the Uniform Building Code discussed in Chapter 7?

The ISO-CLM classifications are used to distinguish between those buildings that can be class rated or specifically rated. Class-rated buildings are those structures that can readily be grouped together because of:

- similarity in size (15,000 square feet or less),
- construction (i.e., classified as frame, joisted masonry, and non-combustible),
- occupancy (i.e., the use to which the premises is being put: apartment units, offices, mercantile purposes, mercantile and apartments or offices, manufacturing, etc.).

Specific rates are reserved for large, high-rise buildings (i.e., above six stories), and buildings with sprinkler systems. Buildings classified as masonry noncombustible, modified fire resistive, and fire resistive must be specifically rated. The specific rate is developed after a detailed inspection of the building. In an effort to contain underwriting costs and make policies more affordable, insurers are moving away from specifically rating commercial buildings, and are making greater use of class rates.

Occupancy

The occupancy of a building determines not only the rate to be charged the individual or firm using or leasing the premises, but also affects the rate applying to the building. For instance, an apartment or office building presents less potential for loss than a building that houses a restaurant, auto repair shop, solvent, or explosives manufacturer.

Because most commercial buildings are not owner-occupied, and since the lessees change from time to time, the underwriter must obtain current occupancy information to evaluate loss potential, and properly rate both building and contents. The type of occupancy will determine whether a building and its contents are eligible for class rates or must be specifically rated. An apartment building is always eligible for class rates, but a mercantile-apartment building with a restaurant on the first floor must be specifically rated. Restaurants, or the buildings housing them, are never eligible for class rates.

Commercial property insurance policies generally provide that an increase of hazard—i.e., a physical change in risk—will suspend or restrict the insurance policy. Thus, the property owner or occupant has an obligation to notify the insurer of changes in any hazard associated with the occupancy.

Location

An important factor in establishing the class or specific rate for a property (building and contents) is its location. In large measure, both class and specific rates are predicated upon a community's fire exposure and experience. Cities and towns are graded for the quality of their fire protection and the availability of their water supply. The Public Protection Classification Manual, developed and published by Insurance Services Offices' Commercial Risk Services, Inc., contains the grading specifics. A grade of 1 means that a community has the best available protection; a grade of 10 means that it has no protection. In classes 9 and 10, listed as "unprotected," the grade is determined by the distance to the closest fire hydrant or fire department; these two classes are generally found in rural areas.

In addition to fire exposure and protection, location is also a factor influencing loss caused by the natural hazards such as hurricanes, tornadoes, severe windstorms, floods, and earthquakes. Over the years, property insurance underwriters have tended not to give significant attention to the location of a risk, since they viewed fire as the predominant cause of loss. With the movement of more and more people into the coastal plains, to the shorelines of the Great Lakes, to the edges of major tributaries, to recreational areas, and to the West Coast, where wildfires and earthquakes pose a threat, a new awareness of natural hazards is emerging among underwriters.

Commercial property insurance underwriters are becoming more knowledgeable about location, and the need for land use planning. Underwriters are not only assessing the physical attributes of structures, but are also weighing the characteristics of the land and the amenities of the location, both of which influence and affect the loss potential.

AVAILABILITY OF DATA ON COMMERCIAL INSURANCE

Little information has been published in the past about the effects of natural hazard perils on commercial structures. Researchers and policymakers are experiencing a growing need for data on claims payments in particular. This information is necessary to identify issues and establish programs for effective, efficient, and economical hazard mitigation. In the past, several factors have worked against the collection and aggregation of loss payment data—namely, competition, antitrust concerns, and privacy laws.

Insurers have not wished to divulge any information to competitors about their insureds. Information on specific risks was deemed proprietary, and was not released for use by others. Another factor inhibiting data collection involves the industry's statistical plans, whose data fields are becoming increasingly complex. The industry has staunchly resisted efforts to expand the statistical plans to accommodate new and desirable information—for instance, a more detailed breakdown of the "cause of loss" codes to distinguish losses caused by hailstorms, tornadoes, hurricane, and other instances of severe winds, rather than designating all of these situations as "wind." The creation of new wind and construction classifications is considered not necessary and extremely costly from an operational standpoint. Use of the statistical plans, it has been argued, should be confined to the acquisition of data elements used in the current rate-making process. This argument does not consider the advent of powerful personal computers and more sophisticated software packages, and the ability to collect and compile data that previously had been beyond reach.

INSURANCE AGAINST NATURAL HAZARD DAMAGE TO SMALL BUSINESSES

Damage to commercial structures caused by natural hazards can affect society as severely as natural hazard damage to residential buildings. In this regard, the small business area particularly requires attention. The paper by Daniel J. Alesch and James N. Holly (1996) provides the following insights:

> Small businesses are particularly vulnerable to earthquakes, hurricanes, and widespread flooding for several reasons. First, they tend to have all their eggs in one basket. That is, they are rarely diversified in terms of the products or services they offer. They usually have only one location from which they do business, and, often, their customers are victims of the same natural disaster. Chances are good that their owner-operators will also suffer damage to their home[s], increasing significantly [their] financial and psychological burden[s]. Finally, small businesses seldom have much political clout, either individually or collectively. With a few notable exceptions, small business owners are not organized and have little success putting pressure on government for special attention or special assistance. (pp. 1–2)

The report goes on to cite the following community impacts which, in turn, provide additional reasons for the need for a comprehensive research program.

> Because small businesses are vulnerable, earthquakes and other natural hazard events often put many of them out of business for durations ranging from a few hours to forever. When small businesses fail— when they go out of business permanently—there are costs to the individual and to the community. The direct monetary costs include destruction of physical capital such as buildings, machinery, and inventory. They include, too, the loss of wages to employees and the loss of income and assets to owner-operators. (p. 2)

Finally, the report emphasizes social costs.

> The social costs of disaster-induced business failure are also important. The community loses retail outlets and services at a time when they may be important to recovery. Not everyone is cut out to be a successful small business entrepreneur so, when some are lost because they are bankrupted by a disaster, new job creation suffers. A more complex consequence of a natural disaster is that the neighborhood "system" is disrupted and that change dynamics may be altered drastically, either accelerated or altered. Fundamental recovery may be delayed for months or even years. One of the most significant social effects may be a significant downward social and economic transformation of the affected neighborhood or community. (p. 2)

The importance of small businesses in the economic scheme of the United States is difficult to overstate and should not be overlooked. In an era of downsizing, small businesses have provided a cushion, absorbing in their employment ranks many thousands of the individuals cut adrift. Alesch and Holly chronicle the plight of small businesses in the face of extreme natural hazard events, based on the effects of the Northridge earthquake. Further research is needed to determine the extent to which their findings apply to the impact of other types of natural disasters, such as floods, hurricanes, and other types of earthquakes. The authors, however, delineate crucial issues that commercial insurers grapple with as they try to provide products that are available and affordable while protecting the solvency of their companies.

Evaluating Models of Risks from Natural Hazards

CRAIG TAYLOR, ERIK VANMARCKE,
AND JIM DAVIS

*A model should be as simple as
possible, but not simpler.*
—Albert Einstein

WHAT ARE MODELS OF RISKS from natural hazards? What functions do they serve? How can they be evaluated and tested? What is the current status of these models? Do they serve the purposes intended? Can we trust them?

These questions arise in a wide variety of contexts, but of special interest are those contexts in which public concern is expressed over the degree of confidence that should be placed in various models of risks from natural hazards and in the results that these models produce. The use of risk models—often with proprietary components—for hurricane and earthquake rate-making often produces heated controversies among scientists, engineers, and the public at large (Dennis Kuzak, EQECAT, personal communication, 1996; Jack E. Nicholson, Chief Operating Officer, Florida Hurricane Catastrophe Fund, personal communication, 1996). For instance, many questions have arisen about the contrast between two rates based on proprietary models: the very low rates established in the original California earthquake insurance program (under the FAIR Plan) and the rates that now appear very high to the public in the new California

Earthquake Authority (CEA) insurance program. (For more details on the CEA, see Chapter 4.) Is there an acceptable level of consistency and validity in how the rates are developed? Is there a sound statistical basis for the development of the rates? Do alternative assumptions used in risk analysis techniques lead to either lower or higher rates? From an insurance perspective, do the risk analysis techniques used integrate with reinsurance pricing and costs? How can the public have confidence in rate-making procedures if these are proprietary? While applying modeling to catastrophe insurance issues has received much publicity, licensing considerations, standardization requirements, cost-benefit studies, risk reduction prioritization procedures, and tools for emergency response planning have also raised significant issues about the application of new risk technologies to social problems.

This appendix first outlines what we mean by models of risks from natural hazards. Then, we identify some of the functions of these models and address issues involved in evaluating and testing these models, such as their trustworthiness considering their proprietary nature when used for public purposes. We do not address in detail how these risk models are constructed, although evaluation and testing of models obviously requires intimate knowledge of this process.

From a public standpoint, evaluation testing and standardization procedures are necessary when the models are employed to set rates or regulations in the public interest. At present no consensus exists on models produced in the private sector. In fact, rigorous evaluations of models have to date produced varying results. CEA hearings in California and activities of the Florida Commission on Hurricane Loss Project Methodology have illustrated the varying results of disciplined queries into current models. (See State Board of Administration of Florida, 1996, for examples.) There are questions at present as to how these evaluation and testing procedures—as well as the development of guidelines, criteria, and standards—can be most effective and equitable in serving public and private needs. Attempts currently under way to evaluate and test proposed risk models could better define their appropriate uses and limitations. To the extent possible, given that proprietary elements in modeling may be legitimate in some cases, these efforts can build successfully on past endeavors to evaluate theories, data, and projections.

MODELS OF RISKS FROM NATURAL HAZARDS:
WHAT ARE THEY?

To respond to this question, we propose some formal characterizations of models of risks from natural hazards, generally consistent with discussions in EERI (1984) and Sauter (1996). These formal characterizations can be supplemented with a historical approach to natural hazards model developments. Models should be understood in relationship to the issues that they are designed to address. Sometimes models focus on very specific practical or theoretical issues, but are then used to address other issues for which they are ill suited.

A *model* is a general representation of reality. The reality represented may be physical, as in the case of a natural hazard, or it may include significant human elements, as in a model of damage to a building or of a financial risk. A model may be used to guide our understanding and action with respect to the phenomena that it represents. Modeling is an activity associated with scientific, engineering, economic, financial, or more generally technical activities. Models can be valuable as tools to place constraints on, say, the dimensions of losses owing to natural hazards. But not everything that we say or think can be expressed by a model. There may not be a model for the salvage value of real estate after a landslide because there is only anecdotal information on the impact of such events on property value. Most landslides, we expect, greatly harm the value of the real estate that they affect. Some landslides may have cleared away old structures which are typically vulnerable to earthquakes and extreme winds, and in rare cases may have stabilized the slope, enabling developers to construct new and less vulnerable buildings in prime real estate areas. This may be worth looking into. However, as long as we rely only on anecdotal and conjectural information, we do not have much of a model.

Models are often primarily *qualitative*. This is true, for instance, of models of political theory, law, organizational behavior, or even models in the fields of biology and geology. In spite of the primarily qualitative nature of these models, they may still be buttressed by quantitative information. Thus, case studies of how organizations function may be based on quantitative data, but the models of organizations constructed from the data may be largely qualitative. Micro-zoning natural hazards for land use planning may be based on qualitative as well as quantitative models, but even the qualitative models are likely to be supported by quantitative information.

Quantitative models vary in the degree to which they are *determin-istic* as opposed to *statistical* or *probabilistic*. Many models in physics are primarily deterministic, in the form of a "law" or assertion that some phenomena are universally valid. A deterministic forecast of the tides at Laguna Beach will, for example, postulate their predicted height at some point of time in the future, and will ignore variations in this estimate. In contrast, statistical or probabilistic models incorporate the notion of variations and possibly the factors that cause or influence these varia-tions.

Models may be deterministic in most respects, but still contain ran-dom variables or functions, or various statistical or probabilistic consid-erations. For instance, some early models of earthquake insurance risks incorporated many statistical elements in order to arrive at a 90th per-centile estimate of a loss from a maximum credible earthquake. These estimates are deterministic insofar as they are based on a single scenario, but they may utilize statistical elements when providing damage ranges. Real-time models of hurricanes, flooding, tornadoes, or earthquakes may likewise be based on a single scenario, but have probabilistic features when quantifying the variability in forecasting the impacts of these deter-ministic events.

Current models of earthquake insurance risks are primarily statisti-cal, yielding probability distributions of losses obtained from a represen-tative ensemble of scenarios. Some early models of hurricane risk began in a similar fashion by modeling loss distributions (Friedman, 1974). Current hurricane models have built on this earlier tradition.

A *model of risks from natural hazards* is a general representation of the chances of damage, disruption, loss, injury, death, or contamination as a result of one or more natural hazards potentially acting on people, structures and other facilities, and/or the environment. A comprehensive model of a risk from a natural hazard generally requires many models and submodels such as models of *exposures* (of persons, human works, the environment), *natural hazards* (their character, distribution, fre-quency, and their local physical effects), *vulnerabilities* (the proneness to varying degrees of loss or dysfunctionality relative to local physical ef-fects or severities), and *risk calculations* (the systematic use of the other component models in order to provide quantitative information on risk).

Modeling risks from natural hazards is actually a specialized field within science and engineering. Most scientists and engineers working in this area focus on some portion of the overall risk problem, such as how to design steel buildings to withstand seismic and wind loads or how to identify and evaluate landslide potential. Even though the history of efforts to model risks from natural hazards is unfortunately sparse, insights into various challenges and developments in risk assessment for natural hazards are provided by Freeman (1932), Friedman (1974), Petak and Atkisson (1982), Steinbrugge (1982), and Eguchi and others (1989).

It is desirable to supplement formal characterizations of models of risks from natural hazards not only with an account of the history of modeling efforts, but also with an understanding of the purposes of the models. Typically, models involving natural hazards are designed to solve one specific issue or set of issues. As a consequence, the models involving natural hazards designed for some limited purpose cannot be readily adapted to other purposes. For instance, models used in the design of resistant buildings, the qualification of nuclear power equipment for licensing, the remediation of soils, or the designation of landslide zones for initial screening purposes, all typically seek to quantify or justify various "safety factors." When such built-in factors of safety or conservatism are part of models used in insurance and reinsurance, practical results may be biased and impact the pricing, availability, and affordability of coverage. Similar remarks apply to the use of design models for regulatory efforts, such as the development of cost-benefit ratios. To understand models for natural hazards, it is thus desirable to understand the specific and limited purposes for which they are used.

WHAT ARE MODELS OF NATURAL HAZARD RISKS USED FOR?

Natural hazard risk models have a large number and variety of uses in insurance and government. They can serve as decision support systems for:

- emergency response planning;
- real-time emergency actions such as evacuation;
- real-time response and recovery activities such as determining federal and state disaster assistance, and relocating displaced households or managing claims adjustment activities;
- pre-disaster land use planning such as avoidance of various hazardous sites or requiring further evaluation and possibly geotechnical remediation for these sites;
- pre-disaster preventive actions, such as the choice of building codes and building code design levels, and their enforcement;
- developing regulatory criteria for assessing whether insurers are

over-committed with respect to their coverages of risks from natural hazards;
- developing regulatory criteria for assessing the suitability of private insurer rates for natural hazards risk coverages;
- evaluating the effectiveness of public insurance programs to cover
  risks from natural hazards;
- developing estimates of pure premiums and loss exceedance probabilities for use in actuarially based premiums and in financial
  planning;
- developing underwriting guidelines or other strategic underwriting measures;
- selecting how much reinsurance to purchase.

These models should be evaluated in relation to the uses for which
they are designed. Those that are adapted for one use may be poorly
adapted for others. For instance, models for evaluating and updating
real-time losses are probably ill-adapted to the evaluation of building
codes and building design levels. Greater attention needs to be paid as to
whether or not the quantitative results of models, however accurate, serve
the needs of the decision makers. The principal function of models of
risks from natural hazards is to systematize and integrate considerable
diverse information into a format that is useful in decision making. The
issue of whether models serve this purpose, and whether the criteria used
are appropriate, remains to be evaluated.

HOW CAN MODELS OF HAZARD RISKS BE
EVALUATED AND TESTED?

The nature of the uncertainties in analyses of risks from natural hazards has long been considered important and now is receiving a great
deal of attention. (See, for example, Abrahamson et al., 1990, for a
provocative discussion of the distinction between random and epistemic
uncertainties; and the National Research Council, 1997, for a critical
review of this distinction. For in-depth discussions of the development
of probability theories, and the philosophical issues to which they give
rise, see Will, 1974.) Currently, concern over the new application of
risk-based models for earthquakes and hurricane insurance rate-making
has led to an increased focus on the question of how accurate they are.
Can these models be used to develop very precise rates? These models
probably cannot "incorporate" all uncertainties in some quantitative

fashion because of their complexity and the uneven development of the technologies that underlie them. The new application of risk tools in rate-making has focused attention on how public officials should respond to the uncertainties inherent in these models.

One characteristic of very complex models, such as those developed for hurricanes and earthquakes, is their uneven development. For some of the submodels, there may be very substantial databases that give them a high degree of reliability. For instance, some underwriters maintain a meticulous record of their exposures based on standardized inspections and the use of standardized policy forms. For some submodels, such as how seismic wave amplitudes decrease as seismic waves move through the earth's crust, substantial databases will reduce but not eliminate the variability in estimates. In other instances, submodels are highly theoretical or subjective, based on almost no data such as the modeling of business interruption or the effects of retroactive building codes on post-disaster repair costs. (See Eguchi et al., 1989; Morrison, 1985; Taylor et al., 1998, for descriptions of the problems of applying linear models to capture systemic impacts, such as business interruption, and the problems in applying oversimplified subjective models of direct damage-ability.) For some submodels, any number of competing approaches may be equally valid, considering the degree of uncertainty surrounding the estimates. To the extent that this is true, models of risks from natural hazards may be more suitable for indicating trends than for providing very precise results.

How then does one evaluate and test natural hazards models when they are in a state of flux and uneven development? For exposition purposes, we first offer some comments on various forms of *testing* and *validation*. We then consider briefly the common method of *corroboration*. Finally, we consider the development of *guidelines, criteria,* and *standards*, which some government officials have in the past resorted to in order to assure that there is a fair playing field and that the performance of facilities subjected to natural hazards can be compared against certain nominal or minimally acceptable standards.

Given the unevenness of information available for developing accurate component models or submodels, one means to guide *testing* and *validation* is to perform *sensitivity analyses*. These analyses indicate whether the uncertainties in submodels make much of a difference to the overall risk estimates produced. These analyses can yield surprising results. Sometimes improving a submodel will not matter much in spite of the limitations of the information on which it is based. For instance, a precise estimate of the salvage value of real estate in a landslide region may not be needed in order to justify in cost-benefit terms a community program to avoid development in landslide-prone regions. Sensitivity analyses can thus serve to focus on those uncertainties that make greater differences to overall model results.

In further preparation for testing and validation, public agencies can themselves gather pertinent data that can be used in models. The United States Geological Survey, state geological agencies, the National Oceanic and Atmospheric Administration, and the National Flood Insurance Program are among the agencies that provide data that can be used in risk assessment for natural hazards. The California Office of Emergency Services, supported indirectly through the Federal Emergency Management Agency, has spearheaded efforts to accumulate loss and damage data from the 1994 Northridge earthquake. The California Department of Insurance has taken the lead in compiling earthquake insurance loss data so necessary to the development of earthquake loss models.

Not only do these institutionalized data developments assist in improving existing models, but public data sources can also serve as benchmarks in assessing the validity of models. For example, public data on wind velocities can assist regulators in assessing whether existing hurricane models adequately quantify wind velocities, an important element in the modeling process. By using these data as benchmarks, regulators can thus provide greater assurance to the public that modeling efforts have been critically scrutinized (Jack E. Nicholson, FHCF, personal communication, 1996).

Actual testing and validation procedures are well understood in principle although sometimes complex and difficult to implement. They include quality assurance on all submodels, as well as evaluation of the relevance of the models for the problems at hand. Such *testing* and *validation* can be very complicated and require considerable data and expertise. How, for instance, can one distinguish between a forecast that is lucky (e.g., comes out about right in one event, as a result of offsetting errors) and one that is off by two standard errors as would be expected in some percentage of cases? How can one determine that the statistical and numerical procedures used are sound? How can one show that the submodels are based on the state of the art?

Since testing and validation require considerable expertise, one may be inclined to use *corroboration* as a means of evaluating natural hazards models. In professional circles, the use of peer review is common, and can signify to the general public that at least a second opinion has been sought. Since second opinions can also be flawed or biased, another proposal is the construction of *guidelines, criteria,* or *standards* for the adequacy of specific natural hazards models (Jack E. Nicholson, FHCF, personal communication, 1996; State Board of Administration of Florida, 1996).

Guidelines are suggestions for future action but may eventually be modified into criteria. *Criteria* are products of scientific inquiry, not the result of a process where personal values, negotiation, and compromise guide the results. They are expressions of existing knowledge and may be only feeble representations of the truth when knowledge is incomplete. If properly prepared, criteria should be useful to legislators, policymakers, and citizens as scientifically defensible and unbiased sources of information about what is known.

Criteria represent a general agreement or consensus which should be tied to documents specifying the current state of knowledge or technology. Criteria also serve as a basis for continuing scientific inquiry and research and provide decision makers with information to weigh costs and benefits associated with alternative strategies and objectives (Atkisson and Gaines, 1970; Faith and Atkisson, 1972).

Standards, on the other hand, are the product of policy and political considerations. Whereas criteria are descriptive or predictive, standards are prescriptive in that performance (code) levels must be legally met. The adoption of standards based upon criteria established as a result of codifying the state of knowledge should govern the performance of models (Atkisson and Gaines, 1970; Faith and Atkisson, 1972).

Using guidelines, criteria, or standards overcomes the ad hoc limitations of validation and testing on a case-by-case basis. One knotty problem currently faced by regulators is how to assess proprietary hurricane and earthquake models used by insurance companies. Should these models be licensed? Should insurers be allowed to use them for setting specific rates (or even in setting very rough rating guidelines)? At what point and to what extent should regulators and the general public have access to the details of these proprietary models? The use of guidelines, criteria, or standards assists the regulator in addressing these difficult issues by their consideration of both public and private interests. There is some urgency in taking this step when risk models are used to develop homeowners' insurance rates and these homeowners lack confidence in how their rates are determined.

The effectiveness of the use of guidelines, criteria, and standards can be illustrated by the practice of licensing professionals working in the geosciences. (Over half the states have these licensing requirements and guidelines for geologists, engineering geologists, and/or geophysicists.) In California in 1969, registration was required for minimum qualification for practicing geology and geophysics as it impacts safety. Since that time, standards have been refined, and specialties such as engineering, geology, and hydrology have been added. This licensing practice received even more attention after 1989. Some practitioners have decided to cease practice and others have been brought before the Professional Registration Boards.

Another example of the challenges of implementing guidelines or standards is Freddie Mac's recent earthquake requirements for new loans, which has led to a reaction embodied in the signing of Senate Bill No. 1327 (Section 10089.4 of the Insurance Code) in California. This bill in turn has led to questions on how to operationalize the following statutory language:

> No person may use a geographically based earthquake assessment system or program for the purpose of requiring earthquake insurance, or imposing a fee or any other condition in lieu of requiring earthquake insurance, in connection with a loan secured by a condominium project or an individual unit of a condominium project unless the system's or program's analytical assumptions and methodology used in the assessment have been submitted to and approved by the commission. In determining whether to approve a submission, the commissioner shall consult with and consider the input of the State Geologist.

The foregoing language confirms that the public requires greater confidence than now exists in the new applications of models of risks from natural hazards to the development of detailed mortgage underwriting guidelines.

The development of criteria, guidelines, and standards thus promises to create a fair playing field in how risk models for natural hazards are used. How urgent it is to develop these criteria, guidelines, and standards depends on the degree to which decisions impacting the general public are being made based on these risk models. If the experience with regulation in the geosciences is any indication, the development of these criteria, guidelines, and standards will take considerable time, with potentially disappointing results, especially during periods when public awareness is diverted to other issues.

SUMMARY

Models of risks from natural hazards are general representations of how disasters can cause damage, loss, injury, death, or environmental impairment. As such, these models, if tested and validated, can serve to guide a variety of public and private actions in insurance and government. Recent public concern over the uses of some of these models raises the issue of how public officials should respond. In particular, what ways will effectively assure that public interests are served without interfering with private markets?

Questions about testing, validation, guidelines, criteria, and standards for natural hazards that have clear public impacts are significant because there is in general no current consensus on natural hazards models. Various attempts to test models in terms of their prediction of actual events, through simulations and their scientific and engineering bases, have not yielded any consensus. One reason for this has been the very complexity of the models themselves, some of which could, without harming available evidence, be replaced by any number of competing quantitative models. Private vendors have also been reluctant to divulge trade secrets.

The state of knowledge, which is the basis for specifying criteria, must be open and available. The criteria-setting process, in turn, is the basis for the development of standards that in the judgment of government administrators allow for appropriate margins of safety and public welfare. Recent rate hearings have resulted in a compromise that proposed rate increases be supported by adequate evidence for the increase. This compromise has not, to date, required full disclosure of private models. Even more recently, in S.B. 1327 the California State Geologist must be conferred with on mortgage-lending requirements based on private models.

These processes are in a state of flux. As a result, we must look to the future for the development of guidelines, criteria, and standards and await further tests to determine the validity of the many models that exist to assess risks from natural hazards. Ongoing public efforts, however, could continue to illuminate the appropriate uses and limitations of these existing models.

Book Links to the World Wide Web

CHARLES NYCE

R EGARDLESS OF YOUR FIELD of endeavor, one thing always remains constant: time brings change. In the insurance industry, the risks that people face, as well as the products and services that insurers offer to handle these risks, are constantly changing. Staying up-to-date is a challenge.

This book is designed as a *current* assessment of research into and applications of insurance related to natural hazards. Today an assessment that remains relevant and timely must be linked to technology. Technology, in the form of the Internet or World Wide Web (WWW), enables the reader to quickly gather up-to-date information within a matter of minutes. This appendix presents links of the topics covered in this book to sources of information on the World Wide Web in order to allow readers the advantage of the continuous updating of information. To use an analogy, this book in print represents a snapshot of the current role of insurance in dealing with natural hazards. However, over time, a photograph becomes dated. Links to the World Wide Web allow readers access to a source of information that is constantly changing and growing.

By linking this book to the World Wide Web, readers can directly interface with current information. The Internet provides an unimaginable amount of information, and sorting through it can be daunting. The links listed here can provide the reader with in-depth or more detailed knowledge about topics discussed in this book. The web pages cited are maintained by organizations that have a strong interest in natural hazards and strategies for dealing with them.

For those readers not already familiar with the World Wide Web, the Internet is an electronic linking of millions of sites containing information on every conceivable topic. To access these sites, the user basically needs three things: a computer, a modem, and a web interface. (For more information on the use of the Web, see *10 Minute Guide to the Internet* by Galen Grimes and Rick Bolton (Que Education and Training, 1997, third edition, ISBN 0789714051); *50 Fun Ways to Internet* by Alan Hoffman (Career Press, 1998, second edition, ISBN 156414322); or *The Internet and the World Wide Web: A Time-Saving Guide for New Users* by Mark Kressin (Prentice-Hall Computer Books, 1997, ISBN 0134937430.)

This appendix contains over 90 links to various web sites, listed in the order in which these topics appear in the book. Some of the more relevant web sites are briefly discussed.

INSURANCE

From supply to demand, and from regulation to insurability, all aspects of insurance as it relates to natural hazards are discussed in this book. There are literally thousands of web sites that deal with the insurance industry. The Insurance Industry Internet Network and the Insurance News Network provide some of the most comprehensive coverage. The National Association of Insurance Commissioners is the best site on the web for regulatory issues.

NATIONAL FLOOD INSURANCE PROGRAM (NFIP)

The National Flood Insurance Program was established in 1968 for the purpose of reducing flood losses and disaster relief costs. The NFIP was developed in response to the private insurance markets' contention that the flood risk was uninsurable. Today, the NFIP is often used as an example of the role the federal government can play in insuring against natural hazards. It has been a pioneer in other related areas such as floodplain mapping and management and the development of the Community Rating System (CRS).

FEDERAL EMERGENCY MANAGEMENT AGENCY (FEMA)

FEMA is the principal government agency dealing with natural hazards. Founded in 1979, FEMA's mission is "to reduce loss of life and property and protect our nation's critical infrastructure from all types of hazards through a comprehensive, risk-based, emergency management program of mitigation, preparedness, response and recovery." FEMA's web site deals with all aspects of natural hazards, from mitigation and emergency preparedness training, to storm tracking, to aid following a disaster.

CONGRESSIONAL MEASURES

There are numerous sites on the web that cover congressional hearings and legislation. This appendix lists three key sites under "Congressional measures for dealing with natural hazards" that allow the reader to track not only natural hazards bills that Congress enacts, but also congressional hearings and measures that do not become laws. This allows the reader to ascertain the attitude of policymakers toward natural disasters. All three sites contain searchable databases on all aspects of the U.S. Congress.

ALTERNATIVE RISK FINANCING

The future of risk financing for catastrophic disasters is likely to be linked to the capital markets. Capital markets in the United States are better able to absorb multi-billion dollar losses than are private insurance markets. Property Claims Service (PCS) options offered by the Chicago Board of Trade (CBOT) and Act of God bond issues are the first steps to securitization of catastrophe risk. Finding online information in this rapidly developing field has not been easy. Aside from the CBOT, there are no comprehensive web pages that deal with these topics.

Many states have attempted to circumvent problems in the private homeowners' insurance markets by developing residual or windstorm pools. The state of Florida is just one example. The Florida Windstorm Underwriting Association, or FWUA, provides wind coverage to property owners living on the coast in Florida who cannot obtain wind coverage from private insurance companies. Created in 1970, the FWUA was designed to cover wind risk in the Florida Keys, but has since expanded to 29 of the 35 coastal counties in Florida.

INDIVIDUAL NATURAL HAZARDS

There are numerous web sites that deal solely with individual hazards and their impacts. EQE International, FEMA, and U.S. Geological Survey are all organizations that provide fairly intensive web sites dealing with earthquake, hail, hurricanes, landslides, lightning, tornadoes, wildfires, winter storms, and volcanoes. Both EQE International and FEMA provide sites that have storm-tracking capabilities for up-to-the-minute information on the latest storms.

OTHER NATURAL HAZARD SITES

The Natural Hazards Center at the University of Colorado, through its web site, provides interested parties with a wealth of information regarding all aspects of natural hazards. An index page regarding every type of natural hazard encountered in this country is available, as well as information on publications regarding natural hazards. This index provides the reader with an excellent starting point for interactive research regarding natural hazards.

Disaster Finder. This web site, maintained by NASA's Goddard Space Flight Center, "offers a complete index to the best disaster web sites on the Internet" and has won several awards for "Best of the Net." It provides the reader with the option to search for key words, or to browse through the Disaster Finder's own categorization of related sites. Some of the categories covered by this site include general hazard information, disaster management, and organizations concerned with natural hazards.

INDEX OF LINKED WEB SITES

Topics	Chapters	Web Site
Indexes of web sources related to natural hazards		http://ltpwww.gsfc.nasa.gov/ndrd/disaster http://www.colorado.edu/hazards/sites/sites.html
FEMA/mitigation, emergency training, disaster aid	1,6,7	http://www.fema.gov http://www.disasterrelief.org
Migration patterns in U.S.	1,5	http://www.census.gov/population/www/socdemo/migrate.html
Congressional measures dealing with natural hazards	1,2,4	http://www.access.gpo.gov/nara/cfr/cfr-table-search.html http://www.law.vill.edu/fed-agency/fedwebloc.html http://thomas.loc.gov
National Flood Insurance Program (NFIP)	1,2,6,7	http://www.nrc.state.ne.us/floodplain/flood/nfip.html http://www.nrc.state.ne.us/floodplain/flood/flood1.html http://www.fema.gov/nfip
Standard homeowners' insurance policies	1,2	http://www.insure.com/home/index.html
Lender mortgage requirements	1,3	http://www.fanniemae.com http://www.freddiemac.com
Construction / building codes / mitigation	1,7	http://www.fema.gov http://www.cabo.org/index.html/ http://www.buildingcodeonline.com
Stafford Disaster Relief and Emergency Assistance Act	1,6	http://www.fema.gov/home/laws/stafact.htm
Small Business Administration (SBA) disaster loan program	1,6	http://www.sbaonline.sba.gov
Information technology (IT)	1,9	http://www.csd.uu.se/~ae96a1er/essay.html
National Association of Insurance Commissioners (NAIC)	1,8	http://www.naic.org
McCarran-Ferguson Act	1,8	http://www.law.cornell.edu/uscode/15/ch20.html

continued

INDEX OF LINKED WEB SITES—*Continued*

Topics	Chapters	Web Site
Hazard risk maps	2,3,4	http://www.consrv.ca.gov/dmg/index.htm http://www.usgs.gov/themes/FS-248-96 http://iiaa.iix.com/ndcmap.htm
Catastrophe modeling	2,5,7,9	http://www.riskinc.com/rms http://www.eqe.com
Natural hazards - earthquake, hail hurricane, landslides, lightning, tornadoes, wildfires, winter storms, volcanoes	2,3,4,5	http://www.eqe.com http://www.usgs.gov/themes/hazard.html http://www.Colorado.EDU/hazards http://quake.usgs.gov
Bermuda insurance markets	2,5	http://www.bermuda-online.com/insurebm.htm
Fair Access to Insurance Requirements (FAIR) plans	2	http://www.law.cornell.edu/uscode/12/1749bbb-3.shtml http://photon.indy.net/ifp/fairint.htm
California Earthquake Authority (CEA)	2,4,9	http://www.insure.com/states/ca/home/ceabkg.html
Texas Hawaii Florida	2,5,8	http://www.tdi.state.tx.us http://www.hawaii.gov http://www.doi.state.fl.us/index.htm
Hawaii and Texas windstorms	2,5,9	http://www.insure.com/home/windstorm.html
Lloyd's of London	2,6,8	http://www.lloydsoflondon.co.uk
Regulation of insurance	2,8	http://www.naic.org/geninfo/about/regutra3.htm
Northridge earthquake	2,3,4	http://geohazards.cr.usgs.gov/northridge
California Insurance Dept.	2,3,4	http://www.insurance.ca.gov
Capacity and Probable Maximum Loss (PML)	2,4,5	http://www.insurance.ca.gov/FED/EARTHQ97.htm
Alquist-Priolo Special Studies Zones Act	3	http://www.consrv.ca.gov/dmg/ghap/ap-intro.htm
Earthquake insurance disclosure	3,4	http://www.clta.org/LEG/LEGSUM95/e-j.html#ceabeip

INDEX OF LINKED WEB SITES—*Continued*

Topics	Chapters	Web Site
California insurance code	3,4	http://www.ca.gov/s/govt/govcode.html
California Residential Earthquake Recovery Fund	3,4	http://www.consumerwatchdog.org/public_hts/earth/cea/bills/ab13.htm
Homeowners' Guide to Earthquake Safety	3	www.seismic.ca.gov
Federal Insurance Administration (FIA)	3,6,7	http://www.fema.gov/nfip
National Flood Insurance Reform Act of 1994	3,6	http://www.state.in.us/acin/dnr/water/juldec95/answers.html
Applied Technology Council	4,7	http://www.ngdc.noaa.gov/seg/hazard/resource/bldgtech.html
Joint Underwriting Association (JUA)	5	http://www.insure.com/states/fl/juamain.html
Florida Hurricane Catastrophe Fund	5,9	http://www.pubadm.fsu.edu/collins/programs/economic/hurricane/equation.html
Florida Windstorm Underwriting Assoc. (FWUA)	5	http://www.fwua.com/index.html
Disaster planning	5,7, Appendix B	http://www.ul.cs.cmu.edu/books/mitigating_losses/miti001.htm http://www.fema.gov
State guaranty funds	5,8	http://www.naic.org/geninfo/about/regutra3.htm
Generally accepted accounting principles (GAAP)	5,8	http://badger.ac.brocku.ca/~kr96ad/gaap.html
A.M. Best and Co.	5,8	http://www.ambest.com
Standard and Poor's	5,8	http://www.ratings.standardpoor.com
National Oceanic and Atmospheric Administration (NOAA)	5, Appendix B	http://www.noaa.gov

continued

INDEX OF LINKED WEB SITES—*Continued*

Topics	Chapters	Web Site
Institute for Business and Home Safety	6,9	http://www.iiplr.org/index.htm
Dept. of Housing and Urban Development (HUD)	6	http://www.hud.gov
Flood Insurance Rate Map	6	http://www.awod.com/unisun/mapinfo.html
Floodplain management, Community Rating System (CRS)	6,7	http://www.usace.army.mil/ncd/fpma.htm
Natural Hazards Center, University of Colorado	6	http://www.Colorado.EDU/hazards
Insurance Services Office (ISO)	6,8, 9 Appendix A	http://iso.com
Erosion	6	http://soils.ecn.purdue.edu/~wepp/nserl.html http://sparky.nce.usace.army.mil/hes/gallery.html
North Carolina, erosion	6	http://www2.ncsu.edu/eos/service/bae/www/programs/ extension/publicat/wqwm/130/index.html http://www.sips.state.nc.us/DOI
North Carolina Division of Coastal Management	6	http://www.nos.noaa.gov/ocrm/czm/ czmnorthcarolina.html
Coastal Barrier Resources System	6	http://www.netlobby.com/coastal.html
Dept. of the Interior - ecological integrity	6	http://www.ios.doi.gov/nrl
Interagency Floodplain Management Review Committee	6	http://www.usace.army.mil/ncd/fpma.htm
Risk-based capital (RBC)	8	http://www.captive.com/newsstand/articles/13.html
Captives, risk retention groups (RRG)	8	http://www.captive.com
Chicago Board of Trade (CBOT)	Appendix A, 9	http://www.cbot.com

INDEX OF LINKED WEB SITES—*Continued*

Topics	Chapters	Web Site
Act of God bonds	9	http://www.artn.nwu.edu/Richard/tools.html
Insurance Information Institute (III)	Appendix B	http://www.iii.org
California Office of Emergency Services	Appendix B	http://www.oes.ca.gov

Glossary[1]

Actuarially sound rates. Rates that accurately reflect actual loss experience, as opposed to rates that are subsidized or are inadequate.

Admitted company. An insurance company authorized and licensed to do business in a given state. (Gloss.)[2]

Affordability. Usually applied to those cases where a person or a group of persons do not have the money to purchase insurance, or as much insurance as they would like to have; often with the implication that rates are simply too high and subsidies may be necessary.

---

[1]This glossary was prepared by Robert DeChant.

[2]Key to Sources:

(Encyc.) = Encyclopedia of Associations, Sandra Jaszczak, editor. Detroit: Gale, an International Thomson Publishing Co., 1996.

(Gloss.) = Glossary of Insurance Terms. Santa Monica, Calif.: The Merritt Company, 1980.

(I.R.M. Glossary) = Insurance and Risk Management Glossary, Richard V. Rupp, CPCU. Chatsworth, Calif.: Nils Publishing Co., 1991.

(New World) = Webster's New World Dictionary, Second College Edition, 1980.

AIA. American Insurance Association. Merged with the National Board of Fire Underwriters and the Association of Casualty and Surety Companies. Represents companies providing property and casualty insurance and suretyship. Monitors and reports on economic, political, and social trends; serves as clearinghouse for ideas, advice, and technical information. (Encyc.)

Alien Insurer. An insurer formed under the laws of a country other than the United States. A U.S. company selling in other countries is also an alien insurer. (Gloss.)

Availability. Ability and willingness of the insurance industry to provide coverage. Especially applicable to high-risk areas.

BCEGS. Building Code Effectiveness Grading Schedule. A classification of communities by the Insurance Services Office based on how well they have implemented and enforced building codes in their community.

BOACA. Building Officials and Code Administrators International. Government officials and agencies and other interests concerned with administering or formulating regulations for buildings, fire, mechanical, plumbing, zoning, and housing. (Encyc.)

Capacity. The total limit of liability that an insurance or reinsurance company, or the industry, can assume, according to generally accepted criteria for solvency. Can be viewed on countrywide or territorial basis.

Cataclysm. Any great upheaval that causes sudden and violent changes, as an earthquake, war, great flood, etc. (New World)

Catastrophic Risk. The risk of a large loss by reason of the occurrence of a peril to which a very large number of insureds are subject. (Gloss.)

Catastrophic Loss. Damage resulting from a catastrophe.

CATEX. An exchange through which insurers trade "standardized catastrophe units."

CBOT. Chicago Board of Trade. Provides a futures exchange dealing with contracts based on agricultural products, financial instruments, precious metals, and options on futures including catastrophes. (Encyc.)

CEA. California Earthquake Authority, 1996. A state agency, established by state statute, with the authority to act as an insurer of earthquake coverage for residential property.

Claim. A request made by an insured for payment for damages covered by insurance.

Coinsurance clause. A policy provision stating that the insured and the insurer will share all losses covered by the policy in a proportion agreed upon in advance. (Gloss.)

CRERF. California Residential Earthquake Recovery Fund. A program to insure the 10 percent deductible applicable to earthquake insurance, to a maximum of $15,000, via a small premium surcharge. Program created and repealed in 1992.

Cross-subsidies. When insurance is priced below cost for particular geographic areas or groups of policyholders, a portion of their losses over time is subsidized by premiums collected in other areas or from other groups of policyholders.

CRS. Community Rating System. A grading of communities for mitigation efforts beyond those required by the NFIP for its policyholders.

Deductible. The portion of an insured loss to be borne by the policyholder before he or she is entitled to recover from the insurer. (Gloss.)

Disaster. A great or sudden happening that results in loss of life, property, etc. How large a happening has to be in order to be classified as a "disaster" depends upon the perspective of the individual, community, state, or country affected.

Domestic company. An insurer formed under the laws of the state in which insurance is written. (Gloss.)

Earthquake magnitude. Earthquakes are measured in terms of the total amount of energy released when slippage occurs along an earthquake "fault" (a break or fracture in the earth's rock strata). The energy release creates vibrations that can be measured by a seismograph. The seismograph readings are described in terms of magnitude numbers based on the Richter scale. Major damage to buildings typically begins to occur when an earthquake's magnitude measures higher than 6 on the Richter scale. The strongest earthquakes ever recorded measured between 8 and 9.

Emergency Program. A temporary installation of the NFIP, with restricted coverage, in a community seeking to qualify for the permanent regular program.

Exposure. 1. The state of being subject to the possibility of loss. 2. The extent of risk as measured by payroll, gate receipts, area, or other standards. 3. The possibility of loss to a risk being caused by its surroundings. 4. Surroundings producing a loss to the insured property. (Note: an example of 3 and 4 is an insured building that suffers loss from a nearby dynamite factory explosion.) (Gloss.)

FAIR Plan. Fair Access to Insurance Requirements. A pooling arrangement to insure risks that cannot obtain coverage in the private market.

FEMA. Federal Emergency Management Agency. A cabinet-level agency responsible for developing mitigation strategies, coordinating disaster relief, and administering the NFIP.

FHCF. Florida Hurricane Catastrophe Fund A tax-exempt trust fund authorized in 1993 which is financed by premiums from insurers based on their property exposure in the state. The fund covers a portion of the losses to insurers from catastrophic disasters to reduce their chances of insolvency. A retention level is specified for each year, and insurers are reimbursed for losses in excess of that level. In the event the fund does not have enough money to pay in full, a set of rules are in place defining the sharing of the funds available.

FIA. Federal Insurance Administration. An arm of FEMA, which administers the NFIP.

FIGA. Florida Insurance Guaranty Fund. Created in 1970 to cover the claims payments of insolvent insurers. FIGA is funded by assessing the property insurance premiums written by all insurers in the state based on their percentage of market share.

FIRM. Flood Insurance Rate Map. Delineates boundaries of 100-year floodplain (1 percent annual probability of flood) in a flood-prone community (see NFIP Manual).

FPCJUA. Florida Property Casualty Joint Underwriting Association. Created by the Florida legislature in 1986 to provide coverage for commercial property that could not obtain insurance in the voluntary market. Expanded in 1993 to accommodate private residential policies of insolvent insurers. (See RPCJUA.)

FWUA. Florida Windstorm Underwriting Association Created in 1970 to provide insurance to individuals who are unable to obtain coverage in the voluntary market for wind and hail coverage in high-wind-risk areas.

GAO. General Accounting Office. A congressionally created watchdog agency for auditing federal government agencies and programs. The GAO also is called upon frequently to assess whether proposed budgets are balanced, to estimate the financial effects of proposed legislation, and to perform other similar tasks.

Hazard. A specific situation that increases the probability of the occurrence of loss arising from a peril, e.g., a flood or earthquake, or that may influence the extent of the loss, e.g., a poorly constructed house. (See Risk.)

Homeowners' insurance. A comprehensive insurance policy covering an owner-occupied dwelling for most physical perils, liability, and theft.

IBHS. Institute for Building and Home Safety. An organization funded by insurance companies, with the mission of reducing deaths, injuries, property damage, economic losses, and human suffering caused by natural disasters. Formerly known as the Institute for Property Loss Reduction (IIPLR).

IFMRC. Interagency Floodplain Management Review Committee. Established in 1994 to study the causes and consequences of the 1993 Midwest flooding.

IIPLR. Insurance Institute for Property Loss Reduction. Name changed to the Institute for Business and Home Safety (IBHS) in 1997. (See IBHS.)

Insurable risk. A risk that an insurer believes it can provide with coverage against future losses.

IRC. Insurance Research Council. Organized to respond to public and insurance industry needs for timely and reliable research findings concerning property and casualty insurance. (Encyc.)

ISO. Insurance Services Office. Seeks to make available to any property and liability insurer, on a voluntary basis, statistical, actuarial, policy forms, and other related services; functions as an insurance advisory organization and statistical agent. Also provides insurance services to the federal government. (Encyc.)

IT. Information technology.

JUA. Joint underwriting association. A generic term used with a number of specific state residual market plans. A pooling arrangement among companies who share risks that cannot be accommodated by a sole insurer in the private market.

Law of Large Numbers. For a series of independent and identically distributed random variables, the variance of the average decreases as the number of items in the sample increases.

Loss. 1. The amount of a reduction in the value of an insured's property caused by an insured peril. 2. The amount sought by an insured's claim. 3. The amount paid on behalf of an insured under an insurance contract.

McCarran-Ferguson Act. Popular name of Public Law 15, enacted in 1945 in response to a Supreme Court decision declaring insurance to be "interstate commerce," and therefore subject to federal antitrust laws. P.L.15 gave insurers a partial exemption from federal antitrust enforcement, to the extent that they are regulated by the state. The law recognizes that insurers need to share loss statistics and engage in joint ventures in order to spread risk and meet the coverage needs of large populations and industrial enterprises. Although the law itself has not been repealed, subsequent court decisions have greatly reduced the list of insurer activities exempt from federal antitrust enforcement.

Mega-Catastrophe. A catastrophe that exceeds the capacity of the industry to respond completely.

Mitigation. Measures taken to reduce or eliminate loss or the impact of events. These measures may be structural, strategic, or educational.

Moral Hazard. Loss to an insurance company that may arise from the dishonesty or imprudence of the insured. (New World) Unethical or careless behavior engaged in by an individual because the property is insured; this behavior increases the probability of loss from an insured peril. Setting a house on fire (arson) as a way of collecting an insurance claim is an extreme example of moral hazard.

NAHB. National Association of Home Builders of the United States. An organization that lobbies on behalf of the housing industry and conducts public affairs activities to increase public understanding of housing and the economy. Collects and disseminates data on current developments in home building and home builders' plans. (Encyc.)

NAIC. National Association of Insurance Commissioners. Association of state insurance regulators responsible for regulating insurance in their respective states. Promotes uniformity of legislation and regulation affecting insurance to protect interest of policyholders. Conducts educational programs for insurance regulators. (I.R.M. Glossary)

NAS. National Academy of Sciences. A private, nonprofit, self-perpetuating society of distinguished scholars engaged in scientific and engineering research, dedicated to the furtherance of science and technology and to their use for the general welfare. Under the authority of the charter granted to it by Congress in 1863, the Academy's working mandate calls on it to advise the federal government on scientific and technical matters.

Natural disaster. An event produced by nature rather than man, i.e., earthquake, flood, hurricane, tornado, volcanic eruption, etc. Sometimes referred to as an "act of God."

NBFU. National Board of Fire Underwriters. A now-defunct organization founded by fire insurance underwriters in 1866. Its main objectives were to promote fire prevention and property loss control. The NBFU was instrumental in developing the standard fire insurance policy in the mid-1960s. The board was merged into the American Insurance Association in the 1970s (see AIA). (I.R.M. Glossary)

NCOIL. National Conference of Insurance Legislators. Made up of chairpersons and members of state insurance or insurance-related committees in state legislatures. Works to support legislators by giving them information to assist them in drafting legislation regulating the insurance industry. (I.R.M. Glossary)

NCPI. National Committee on Property Insurance. Predecessor of the Insurance Institute for Property Loss Reduction (see IIPLR).

NFIF. National Flood Insurance Fund, 1968. A repository for premiums from the NFIP, and from which are paid losses, operating costs, and administrative expenses of the NFIP.

NFIP. National Flood Insurance Program, 1968. Administered by the Federal Insurance Administration (FIA). Provides flood insurance to residents of flood-prone communities that have adopted appropriate land use regulations and building codes.

NOAA. National Oceanic and Atmospheric Administration. Agency established as part of the Department of Commerce in 1970. The mission responsibilities of NOAA are to monitor and predict the state of the solid earth, the oceans and their living resources, the atmosphere, and the space environment of the earth, and to assess the socioeconomic impact of natural and technological changes on the environment. What was formerly known as the U.S. Weather Bureau is now part of NOAA.

Nonadmitted (unauthorized or unlicensed) insurer. An insurer not licensed to do business in the jurisdiction in question. (Gloss.)

Nondomestic insurer. An insurer not domiciled in the state.

NSF. National Science Foundation. Independent agency in the Executive Branch. Concerned primarily with supporting basic and applied research and education in the sciences and engineering. Funds scientific research. (Encyc.)

PML. Probable Maximum Loss. The largest amount of loss likely to occur for a single structure or a geographic area when a given event (e.g., an earthquake) occurs.

Pre-FIRM. In the National Flood Insurance Program, refers to structures constructed prior to the development of a Flood Insurance Rate Map (FIRM) for a community. For these structures rates are subsidized.

Premium. The price of insurance protection for a specified period of time. (Gloss.) Also, the product of rate times amount of insurance.

Probability. Likelihood of an event occurring; the number of times something will probably occur over the range of possible occurrences, expressed as a ratio. (New World)

Rate. The cost per unit of insurance purchased. Usually reflects expected losses from a series of events combined with loss adjusting expense, overhead, and profit.

Regular Program. The phase of a community's participation in the NFIP where more comprehensive floodplain management requirements are imposed, and higher amounts of insurance are available based upon risk zones and elevations determined in a flood insurance study. The rates are actuarially based and not subsidized.

Reinsurance. Insurance purchased by insurance company A from insurance company B for the purpose of spreading risk and reducing company A's losses from catastrophes.

Risk. A person or thing insured; the subject of insurance, i.e., a building; the chance the building could be subject to damage or loss, i.e., "is at risk"; uncertainty of loss.

Risk averse. Wishing to avoid risk. The attitude of people whose concern with a risk associated with an investment or business venture is such that they are willing to pay proportionately more than the expected loss. In relation to catastrophic natural disaster, "risk-averse" insurers wish to avoid exposing their company to an unreasonable chance of insolvency and are thus unwilling to take on more risk than their financial capacity can easily handle.

RMMs. Residual Market Mechanisms. Measures designed to provide coverage for those unable to obtain coverage in the voluntary market, e.g., assigned-risk plans, windstorm pools, JUAs, and reinsurance pools.

RPCJUA. Residential Property and Casualty Joint Underwriting Association, 1986. For commercial property. Expanded in 1993 to accommodate private residential policies of insolvent insurers.

SBA. Small Business Administration. Agency that distributes some forms of federal disaster relief to homeowners and businesses, in addition to its primary role of providing loans to small businesses.

SBCCI. Southern Building Code Congress International, Inc. Seeks to develop, maintain, and promote adoption of the standard building, gas, plumbing, mechanical, fire protection, and housing codes. Members include state, county, municipal, or other government entities. (Encyc.)

Stafford (Robert T.) Disaster Relief and Emergency Assistance Act.
Passed in 1988 to provide federal disaster relief funds for repair of
public facilities. (Encyc.)

Standard fire policy (New York Standard Fire Policy). Basic fire and
lightning policy used universally by the industry in the early part of
this century.

Tsunamis. Huge ocean waves usually resulting from Pacific hurricanes,
earthquakes, and volcanic eruptions.

USGS. U.S. Geological Survey. Provides reliable, impartial information
to describe and understand the earth. This information is used to
minimize loss of life and property from natural disasters; manage
water, biological, energy, and mineral resources; enhance and protect quality of life; and contribute to wise economic and physical
development.

Wind Pools. Pooling arrangements similar to FAIR plans and JUAs.

WYO. Write Your Own Program. The program under which private
insurance companies sell and service flood insurance for the NFIP.
The private insurance companies assume no risk but are reimbursed
for their expenses.

Bibliography

A. M. Best and Co. 1992. Best's Insurance Reports, Property/Casualty, 1992 Edition. Rahway, N.J.: A. M. Best.

Abrahamson, N. A., P. G. Somerville, and C. A. Cornell. 1990. Uncertainty in strong motion predictions. In Proceedings of the Fourth National Conference on Earthquake Engineering. Palm Springs, Calif.: Earthquake Engineering Research Institute.

Academic Task Force. 1995. Preliminary Report of the Academic Task Force on Hurricane Catastrophe Insurance: Restoring Florida's Paradise. Miami, Fla.: The Collins Center for Public Policy.

Adams, Michael H. 1996. Florida HO insurers face $22.8 million assessment. The National Underwriter, September 2, p. 6.

Adams, Michael. 1997. 62.4% rate hike sought in Florida. The National Underwriter, Property/Casualty Edition, September 1, p. 1.

Advisory Commission on Intergovernmental Relations. 1992. State Solvency Regulation of Property-Casualty and Life Insurance Companies. Washington, D.C.: ACIR.

Alesch, Daniel J., and James N. Holly. 1996. How to survive the next natural disaster: Lessons for small business from Northridge victims and survivors. Paper presented at the Pan Pacific Hazards Conference, Vancouver, British Columbia, July 29–August 2.

271

American Insurance Association and Alliance of American Insurers. 1995. New Partnerships for Catastrophe Response. Washington, D.C.: AIA.

Anesaki, Masaharu. 1930. History of Japanese Religion: With Special Reference to the Social and Moral Life of the Nation. London: Kegan Paul, Trench, Trubner.

Atkisson, A. A., and R. S. Gaines. 1970. Development of Air Quality Standards. Columbus, Ohio: Charles E. Merrill.

Bainbridge, John. 1952. Biography of an Idea: The Story of Mutual Fire and Casualty Insurance. Garden City, N.Y.: Doubleday.

Barth, Michael M. 1996. Risk-based capital results from the property-casualty industry. NAIC Research Quarterly 2:17–31.

Barth, Michael M., and Robert W. Klein. 1995. Risk-based capital, risk-based pricing, and regulatory monitoring: A three-leg approach to insurer solvency oversight. Paper presented at the American Risk and Insurance Association 1995 Annual Meeting, Seattle, Wash., August 13–15.

Berger, Lawrence, and Howard Kunreuther. 1994. Safety first and ambiguity. Journal of Actuarial Practice 2: 273–291.

Berliner, Baruch. 1982. Limits of Insurability of Risks. Englewood Cliffs, N.J.: Prentice-Hall.

BestWeek. 1996. Catastrophes: A major paradigm shift for P/C insurers. BestWeek: Property Casualty Supplement, March 25, P/C 1–18.

BestWeek, Property/Casualty Edition. 1996. Florida JUA reports surprisingly small deficits, August 19, p. 6.

Brenner, Lynn. 1993. Handbook for Reporters. New York: Insurance Information Institute.

Briggs, Susan. 1996. Florida extends moratorium for three more years. BestWeek, Property/Casualty Edition, May 13, p. 4.

Burby, Raymond. 1992. Sharing Environmental Risks. Boulder, Colo.: Westview Press.

California Department of Finance. 1995. California Statistical Abstract (annual report). Sacramento: the Department.

California Insurance Department. 1975. California Administrative Code, Title 10. (Chapter 5, Sub-Chapter 3, Section 2307) Sacramento, Calif.: CID.

California Insurance Department. 1995. California Earthquake Zoning and Probable Maximum Loss Evaluation Program. Draft. Los Angeles, Calif.: CID.

California Seismic Safety Commission. 1992. The Homeowner's Guide to Earthquake Safety. Sacramento: the Commission.

Camerer, C., and Howard Kunreuther. 1989. Decision processes for low probability events: Policy implications. Journal of Policy Analysis and Management 8:565–592.

Changnon, Stanley A., David Changnon, E. Ray Fosse, Donald C. Hoganson, Richard J. Roth, Sr., and James Totsch. 1996. Impacts and Responses of the Weather Insurance Industry to Recent Weather Extremes. Final Report to the University Corporation for Atmospheric Research. Mahomet, Ill.: Changnon Climatologist.

Cheshire, Ronald H. 1990. The value of a national wind mitigation and engineering bureau. In the National Committee on Property Insurance Annual Proceedings: Expectations and Realities: The Insurance Industry and the International Decade for Natural Hazard Reduction. Boston, Mass.: NCPI.

Cochrane, Hal. 1997. Forecasting the economic impact of a Midwest earthquake. In Economic Consequences of Earthquakes: Preparing for the Unexpected, Barclay G. Jones, ed. Buffalo, N.Y.: National Center for Earthquake Engineering Research.

Cohen, Linda, and Roger Noll. 1981. The economics of building codes to resist seismic structures. Public Policy (Winter):1–29.

Collins Center for Public Policy. 1995. Final Report of the Academic Task Force on Hurricane Catastrophe Insurance. Tallahassee, Fla.: the Center.

Conning and Co. 1994. Lighting Candles in the Wind: Industry Response to the Catastrophe Problem. Hartford, Conn.: Conning.

Covello, Vincent, and Jeryl Mumpower. 1985. Risk analysis and risk management: An historical perspective. Risk Analysis 5:103–120.

Cummins, J. David, and Patricia M. Danzon. 1991. Price shocks and capital flows in liability insurance. In Cycles and Crises in Property/Casualty Insurance: Causes and Implications for Public Policy, J. David Cummins, Scott E. Harrington, and Robert W. Klein, eds. Kansas City, Mo.: NAIC.

Cummins, J. David, and Helyette Geman. 1995. Pricing catastrophe insurance futures and call spreads: An arbitrage approach. The Journal of Fixed Income (March):46–57.

Cummins, J. David, and Mary A. Weiss. 1991. The structure, conduct and regulation of the property-liability insurance industry. In The Financial Condition and Regulation of Insurance Companies, Richard W. Kopcke and Richard E. Randall, eds. Boston: Federal Reserve Bank of Boston.

Cummins, J. David, Scott E. Harrington, and Robert W. Klein. 1995. Insolvency experience, risk-based capital, and prompt corrective action in property-liability insurance. Journal of Banking and Finance 19:511–528.

Dade County Grand Jury. Final Report, Spring Term 1992. Dade County, Fla.: December 14, 1992.

Davidson, Ross J., Jr. 1996. Tax-deductible, pre-event catastrophe reserves. Journal of Insurance Regulation 15:175–190.

DeLollis, Barbara. 1998. Ruling allows FWUA to increase rates. The Miami Herald. March 10, p. N/A.

Doehring, Fred, Iver Deudall, and John M. Williams. 1994. Florida Hurricanes and Tropical Storms, 1871–1993: An Historical Survey. Florida Sea Grant College Program TP-71. Gainesville, Fla.: University of Florida.

Doherty, Neil. 1997. Financial innovation for financing and hedging catastrophe risk. Paper presented at the Fifth Alexander Howden (Australia) Conference on Disaster Insurance, Gold Coast, August.

Doherty, Neil, and Lisa Posey. 1992. Availability crises in insurance markets. Working paper. Philadelphia: Wharton School, University of Pennsylvania.

Doherty, Neil, and S. M. Tinic. 1982. A note on reinsurance under conditions of capital market equilibrium. Journal of Finance 36:949–953.

Doherty, Neil, Anne Kleffner, and Howard Kunreuther. 1991. The impact of a catastrophic earthquake on insurance markets. Prepared for Federal Emergency Management Agency, September.

Doherty, Neil, Anne Kleffner, and Lisa Posey. 1992. Insurance Surplus: Its Function, Its Accumulation and Its Depletion. Boston, Mass.: The Earthquake Project.

Dong, Weimin, Haresh Shah, and Felix Wong. 1996. A rational approach to pricing of catastrophe insurance. Journal of Risk and Uncertainty 12:201–218.

Drabek, Thomas E. 1986. Human System Responses to Disaster: An Inventory of Sociological Findings. New York: Springer-Verlag.

Dunleavy, Jeanne H., Robin Edelman, Daniel J. Ryan, and C. Brett Lawless. 1996. Catastrophes: A major paradigm shift for insurers. BestWeek Property/Casualty Supplement, March 25, p. P/C 2.

EERI Committee on Seismic Risk. 1984. Glossary of terms for probabilistic seismic-risk and hazard analysis. Earthquake Spectra 1(1):39–40.

Earthquake Engineering Research Institute (EERI). 1992. Landers and Big Bear Earthquakes of June 18 and 19. EERI Newsletter, Special Earthquake Report 16(8):1.

Earthquake Engineering Research Institute (EERI). 1994. Northridge Earthquake, January 17, 1994: Preliminary Reconnaissance Report. Report 94-01. Oakland, Calif.: EERI.

Earthquake Engineering Research Institute (EERI). 1995. The Hyogo-Ken Nambu Earthquake, January 17, 1995. Preliminary Reconnaissance Report. Report 95-04. Oakland, Calif.: EERI.

Edwards, W. 1955. The prediction of decisions among bets. Journal of Experimental Psychology 50:201–214.

Eguchi, R., C. Taylor, N. Donovan, S. Werner, R. McGuire, and R. Hayne. 1989. Risk Analysis Methods for a Federal Earthquake Insurance Program. Prepared for the Federal Emergency Management Agency under contract number EMW-88-C-2872. Los Angeles, Calif.: Dames & Moore.

Einhorn, Hillel, and Robin Hogarth. 1985. Ambiguity and uncertainty in probabilistic inference. Psychological Review 92:433–461.

Emerson, Bill, and Ted Stevens. 1995. Natural disasters: A budget time bomb. Washington Post, October 31, p. A13.

Faith,W. L., and A. A. Atkisson. 1972. Air Pollution, 2nd edition. New York: Wiley Interscience Series.

Federal Emergency Management Agency (FEMA). 1990. Loss Reduction Provisions of a Federal Earthquake Insurance Program, Summary. FEMA-201/ September 1990. Washington, D.C.: FEMA.

Federal Emergency Management Agency (FEMA). 1994. Assessment of the State-of-the-Art Earthquake Loss Estimation. Washington, D.C.: National Institution of Building Sciences.

Federal Emergency Management Agency (FEMA). 1995. National Mitigation Strategy: Partnerships for Building Safer Communities. Washington, D.C.: FEMA.

Federal Emergency Management Agency (FEMA). 1997. Report on Costs and Benefits of Natural Hazard Mitigation. Washington, D.C.: FEMA.

Federal Interagency Floodplain Management Task Force. 1992. Floodplain Management in the United States: An Assessment (Report Volumes 1 and 2/FIA-17 and FIA-18). Washington, D.C.: FEMA.

Finefrock, Don. 1995. Florida raises ante for firms taking clients from JUA pool. Journal of Commerce, October 27(412), p. 10A.

Five Foundations for a Better Built Environment. 1997. Report of the Governor's Building Codes Study Commission, December. Tallahassee, Fla.: State of Florida Study Commission.

Florida Commission on Hurricane Loss Projection Methodology. 1996. Report of Activities as of June 3, 1996. Tallahassee, Fla.: the Commission.

Florida Department of Insurance. 1994. Hurricane Andrew Emergency Rules and Hurricane Andrew Emergency Rule Summary. Tallahassee, Fla.: the Department.

Florida House of Representatives, Committee on Insurance. 1993. Final Bill Analysis and Economic Impact Statement, Chapter # 93-409, Laws of Florida. Tallahassee, Fla.: Florida House of Representatives.

Freeman, John R. 1932. Earthquake Damage and Earthquake Insurance. New York: McGraw-Hill.

Freeman, Paul K., and Howard Kunreuther. 1997. Managing Environmental Risks Through Insurance. Washington, D.C.: American Enterprise Institute, and Norwell, Mass.: Kluwer.

French, Steven, and Gary Rudholm. 1990. Damage to public property in the Whittier Narrows earthquake: Implications for earthquake insurance. Earthquake Spectra 6:105–123.

Friedman, D. G. 1974. Computer Simulation in Natural Hazard Assessment. Prepared for the Program on Technology, Environment and Man, Institute of Behavioral Science, University of Colorado. Hartford, Conn.: The Travelers Insurance Co.

Friedman, M., and L. J. Savage. 1948. The utility analysis of choices involving risk. Journal of Political Economy 56:279–304.

Froot, Kenneth. 1997. The limited financing of catastrophe risk. Paper presented at the Insurance Leadership Forum, New York, January 15.

Froot, Kenneth, David Scharfstein, and Jeremy Stein. 1994. A framework for risk management. Harvard Business Review, November-December, pp. 91–100.

Gallagher, Tom. 1993. The Florida Department of Insurance: Hurricane Andrew's Impact on Insurance in the State of Florida. Ft. Lauderdale, Fla.: The Florida Department of Insurance.

Grace, Martin F., and Richard D. Phillips. 1996. Measuring the relative efficiency of the production of regulation by the states: An examination of the U.S. insurance regulatory system. Working Paper. Atlanta, Ga.: Georgia State University.

Gray, William M. 1995. Early Analysis (As of 31 October) of the 1995 Atlantic Tropical Cyclone Activity and Verification of Authors' Seasonal Predictions. Fort Collins, Colo.: Department of Atmospheric Science, Colorado State University.

Grazulis, Thomas P. 1993. Significant Tornadoes: 1680–1991. St. Johnsbury, Vt.: The Tornado Project of Environmental Films.

Guy Carpenter and Co. 1997. No surprises in property catastrophe renewals. The Monitor Report, February, 7 pp.

Hanks, Thomas, and C. Allin Cornell. 1994. Probabilistic seismic hazard analysis: A beginner's guide. Proceedings of the Fifth Symposium on Current Issues Related to Nuclear Power Plant Structures, Equipment and Piping. Raleigh: North Carolina State University.

Hanson, Jon S., Robert E. Dineen, and Michael B. Johnson. 1974. Monitoring Competition: A Means of Regulating the Property and Liability Insurance Business. Milwaukee, Wisc.: NAIC.

Harrington, Scott E. 1992. Rate suppression. Journal of Risk and Insurance 59:185–202.

Harrington, Scott, Steven Mann, and Greg Niehaus. 1995. Insurer capital structure decisions and the viability of insurance derivatives. The Journal of Risk and Insurance 3:483–508.

Harris, Harold W., K. C. Mehta, and J. R. McDonald. 1992. Taming tornado alley. Civil Engineering 62(6):77–78.

Heine, S. J., and D. R. Lehman. 1995. Cultural variation in unrealistic optimism: Does the West feel more invulnerable than the East? Journal of Personality and Social Psychology 68:595–607.

Hogarth, Robin M., and Howard Kunreuther. 1995. Decision making under ignorance: Arguing with yourself. Journal of Risk and Uncertainty 10:15–36.

Hunter, J. Robert. 1994. Insuring against natural disasters: Prevention, planning and panaceas. Journal of Insurance Regulation 12:467–485.

Insurance Executive Association. 1952. Report on Floods and Flood Damage. Cited in U.S. Senate Report No. 1313, Federal Disaster Insurance. Washington, D.C.: U.S. Government Printing Office.

Insurance Institute for Property Loss Reduction (IIPLR). 1995. Homes and Hurricanes: Public Opinion Concerning Various Issues Relating to Home Builders, Building Codes and Damage Mitigation. Boston, Mass.: IIPLR.

Insurance Research Council and Insurance Institute of Property Loss Reduction. 1995. Coastal Exposure and Community Protection: Hurricane Andrew's Legacy. Wheaton, Ill.: IRC, and Boston: IIPLR.

Insurance Services Office. 1991. Commercial Lines Manual. New York: ISO Risk Services.

Insurance Services Office. 1994. The Impact of Catastrophes on Property Insurance. New York: ISO.

Insurance Services Office. 1996. Managing Catastrophic Risk. New York: ISO.

Interagency Floodplain Management Review Committee (IFMRC). 1994. Sharing the Challenge: Floodplain Management into the 21st Century. Report to the Administration of the Federal Interagency Floodplain Management Task Force. Washington, D.C.: IFMRC.

Jaffee, Dwight, and Thomas Russell. 1996. Catastrophe insurance, capital markets, and uninsurable risks. Paper presented at the Wharton Financial Institutions Center Conference, "Risk Management in Insurance Firms," Philadelphia, May.

Johnson, Eric, John Hershey, Jacqueline Meszaros, and Howard Kunreuther. 1993. Framing, probability distortions, and insurance decisions. Journal of Risk and Uncertainty 7:35–52.

Joskow, Paul L. 1973. Cartels, competition and regulation in the property-liability insurance industry. Bell Journal of Economics 4:375–427.

KRC Research and Consulting. 1995. Qualitative research report: In-depth interviews with community officials, lenders, realtors, and advisory board committee members. Unpublished paper dated August 24, 1995.

Kahneman, D., and Amos Tversky. 1979. Prospect theory: An analysis of decision under risk. Econometrica 47(2):263–291.

Keillor, Garrison. 1985. Lake Wobegon Days. New York: Viking, 1985.

King, Rawle O. 1993. Insurance Markets After Hurricane Andrew. Washington, D.C.: Congressional Research Service.

King, Stephanie, and Anne Kiremidjian. In press. Use of GIS for earthquake hazard and loss estimation. In Geographic Information Research: Bridging the Atlantic. London: Taylor & Francis.

Klein, Robert W. 1995. Insurance regulation in transition. Journal of Risk and Insurance. 62:363–404.

Klein, Robert W. 1996. Insurance guaranty funds: Issues and prospects. Paper presented at the Competitive Enterprise Institute Conference on Rethinking Insurance Regulation, Washington, D.C., March 8.

Kleindorfer, Paul, and Howard Kunreuther. In press. Challenges facing the insurance industry in managing catastrophic risks. In Solutions in the Financing of Property/Casualty Risks, Kenneth Froot, ed. Chicago, Ill.: University of Chicago Press.

Knowles, Robert G. 1996. Property insurance crunch in Florida may be nearing an end. Journal of Commerce, May 23(413), p. 8A.

Kunreuther, Howard. 1978. Even Noah built an ark. The Wharton Magazine (Summer)28–35.

Kunreuther, Howard. 1989. The role of actuaries and underwriters in insuring ambiguous risks. Risk Analysis 9:319–328.

Kunreuther, Howard. 1996. Mitigating disaster losses through insurance. Journal of Risk and Uncertainty 12:171–187.

Kunreuther, Howard. 1997. Rethinking society's management of catastrophic risks. The Geneva Papers on Risk and Insurance 83:151–176.

Kunreuther, Howard, and Anne E. Kleffner. 1992. Should earthquake mitigation measures be voluntary or required? Journal of Regulatory Economics 4:321–335.

Kunreuther, Howard, et al. 1978. Disaster Insurance Protection: Public Policy Lessons. New York: John Wiley and Sons.

Kunreuther, Howard, Jacqueline Meszaros, Robin Hogarth, and Mark Spranca. 1995. Ambiguity and underwriter decision processes. Journal of Economic Behavior and Organization 26:337–352.

Kusler, J., and L. Larson. 1993. Beyond the Ark: A new approach to United States floodplain management. Environment 35(5):7.

Lanier, Henry Wysham. 1922. A Century of Banking in New York: 1822–1922. New York: George H. Doran.

Launie, J. J., J. Finley Lee, and Norman A. Baglini. 1986. Principles of Property and Liability Underwriting, 3rd edition. Malvern, Penn.: Insurance Institute of America.

Lebra, Takie Sugyyama. 1976. Japanese Patterns of Behavior. Honolulu: University Press of Hawaii.

Lecomte, Eugene L., and Ronald W. Demerjian. 1997. FAIR, Beach & Wind, and Rural Risk Plans. Draft. Boston: IIPLR, ISO.

Lemaire, Jean. 1986. Theorie Mathematique des Assurances. Belgium: Presses Universitaires de Bruxelles.

Lemaire, J., C. Taylor, and C. Tillman. 1993. Models for earthquake insurance and reinsurance evaluations. In Proceedings of the Second International Symposium on Uncertainty Modeling and Analysis. Los Alamitos, Calif.: IEEE Computer Society Press.

Lewis, Christopher, and Lewis Murdock. 1996. The role of government contracts in discretionary reinsurance markets for natural disasters. Journal of Risk and Insurance 63:567–597.

Lilly, Claude C. 1976. A history of insurance regulation in the United States. CPCU Annals 29:99–115.

Litan, Robert E. 1991. A National Earthquake Mitigation and Insurance Plan: Response to Market Failures. Boston, Mass.: The Earthquake Project.

Maloney, Frank E., and Dennis C. Dambley. 1976. The National Flood Insurance Program: A model ordinance for implementation of its land management criteria. Natural Resources Journal 16:665–736.

Manes, Alfred. 1938. Insurance: Facts and Problems. New York: Harper and Brothers.

Markus, H., and S. Kitayama. 1991. Culture and the self: Implications for cognition, emotion, and motivation. Psychological Review 98:224–253.

May, Peter, and Nancy Stark. 1992. Design professions and earthquake policy. Earthquake Spectra 8:115–132.

Mayers, David, and Clifford Smith. 1982. On corporate demand for insurance. Journal of Business 55:281–296.

Mayers, David, and Clifford Smith. 1990. On corporate demand for insurance: Evidence from the reinsurance market. Journal of Business 63:19–40.

Meier, Kenneth J. 1988. The Political Economy of Regulation: The Case of Insurance. Albany, N.Y.: SUNY Press.

Mittler, Elliott. 1993. The Public Policy Response to Hurricane Hugo in South Carolina. Working Paper # 84. Boulder, Colo.: Natural Hazards Research and Applications Information Center, University of Colorado.

Mittler, Elliott, Craig Taylor, and William Petak. 1995. National Earthquake Probabilistic Hazard Mapping Program: Lessons for Knowledge Transfer. Report prepared for U.S. Geological Survey. Washington, D.C.: USGS.

Mooney, Sean. 1995. Presentation to Academic Task Force on Hurricane Insurance. Photocopy.

Morrison, Robert M. 1986. Business Interruption Insurance. Cincinnati, Ohio: National Underwriter Co.

Mosteller, F., and P. Nogee. 1941. An experimental measurement of utility. Journal of Political Economy 59:371–404.

Munch, Patricia, and Dennis E. Smallwood. 1981. Theory of solvency regulation in the property and casualty insurance industry. In Studies in Public Regulation, Gary Fromm, ed. Cambridge, Mass.: MIT Press.

Munich Reinsurance Co. 1997. Topics. Munich: Munich Re.

Myers, David G. 1992. Social Psychology, 4th edition. New York: McGraw-Hill.

National Association of Insurance Commissioners (NAIC). 1997. Insurance Department Resources Report 1996. Kansas City, Mo.: NAIC.

National Association of Insurance Commissioners (NAIC). 1998. Model Laws, Regulations and Guidelines. Kansas City, Mo.: NAIC.

National Committee on Property Insurance (NCPI). 1993. Building Codes: Defining the Industry Role. Proceedings of the Second Symposium, July 27 and 28. Boston, Mass.: Institute for Property Loss Reduction.

National Fire Underwriters. 1949. National Building Code, 1949 edition. New York: NFU.

National Oceanic and Atmospheric Administration, Tropical Prediction Center. 1997. The Deadliest, Costliest and Most Intense U.S. Hurricanes of this Century (and Other Frequently Requested Hurricane Facts). NOAA Technical Memorandum NWS TPC-1, February. Washington, D.C.: U.S. Government Printing Office.

National Research Council, Panel on Seismic Hazard Evaluation. 1997. Recommendations for Probabilistic Seismic Hazard Analysis: Guidance on Uncertainty and the Use of Experts. Washington, D.C.: National Academy Press.

National Underwriter Co. 1995. Fire Casualty and Surety Bulletin, Personal Lines Volume. Misc. Property Exb-1. Cincinnati, Ohio: National Underwriter Co.

National Underwriter Co. 1996. Fla. rulings locks in HO insurers. The National Underwriter, Property/Casualty Edition. November 11, p. 4.

Natural Hazards Research and Applications Information Center (NHRAIC). 1992. Floodplain Management in the United States: An Assessment Report. Prepared for the Federal Interagency Floodplain Management Task Force. Boulder, Colo.: University of Colorado.

O'Brien, James J., Todd S. Richards, and Alan C. Davis. 1996. The effect of El Nino on U.S. land-falling hurricanes. Bulletin of the American Meteorological Society 77(4):773–774.

Office of Technology Assessment. 1995. Reducing Earthquake Losses. Washington, D.C.: U.S. Government Printing Office.

Oviatt, F. C. 1905. Historical study of fire insurance in the United States. Reprinted in Modern Insurance Theory and Education, T. Kailin, ed. Orange, N.J.: Varsity Press, 1966.

Palm, Risa. 1981. Real Estate Agents and Special Studies Zones Disclosure: The Response of California Home Buyers to Earthquake Hazards Information. Monograph #32, Program on Technology, Environment and Man, Institute of Behavioral Science. Boulder, Colo.: University of Colorado.

Palm, Risa. 1990. Natural Hazards: An Integrative Framework for Research and Planning. Baltimore: Johns Hopkins University Press.

Palm, Risa. 1995. Earthquake Insurance: A Longitudinal Study of California Homeowners. Boulder, Colo.: Westview Press.

Palm, Risa, and John Carroll. 1998. Illusions of Safety: Cultural and Earthquake Hazard Response in California and Japan. Boulder, Colo.: Westview Press.

Palm, Risa, and Michael Hodgson. 1992. After a California Earthquake: Attitude and Behavior Change. Chicago: University of Chicago Press.

Palm, Risa, Sallie Marston, Patricia Kellner, David Smith, and Maureen Budetti. 1983. Home Mortgage Lenders, Real Property Appraisers and Earthquake Hazards. Boulder: Institute of Behavioral Science, Program on Environment and Behavior, Monograph # 38.

Palm, Risa, Michael Hodgson, R. Denise Blanchard, and Donald Lyons. 1990. Earthquake Insurance in California: Environmental Policy and Individual Decision Making. Boulder, Colo.: Westview Press.

Pauly, Mark 1968. The economics of moral hazard: Comment. American Economic Review 58:531–536.

Peltzman, Sam. 1976. Towards a more general theory of regulation. Journal of Law and Economics 19:211–240.

Petak, W. J., and A. A. Atkisson. 1982. Natural Hazard Risk Assessment and Public Policy. New York: Springer-Verlag.

Pfeffer, I. 1966. The early history of insurance. In Modern Insurance Theory and Education, T. Kailin, ed. Orange, N.J.: Varsity Press.

Piper, Arthur. 1995. A competitive edge. The Review, Bermuda Special Report. July. p. vii.

PIPSO Reports. 1997. Exhibit D7. Boston, Mass.: Property Insurance Plans Service Office.

Reilly, Francis V. 1990. Testimony Before the Subcommittee on Science, Research, and Technology, House Committee on Science, Space, and Technology, June 26, 1990, Washington, D.C.: U.S. Government Printing Office.

Reinsurance Association of America. No date. Reinsurance Market Responses to U.S. Natural Disaster Threats. Section 1. Washington, D.C.: Reinsurance Association of America.

Rejda, George. 1982. Principles of Insurance. Glenview, Ill: Scott, Foresman.

Risk Management Solutions. 1995a. What if a Major Earthquake Strikes the Los Angeles Area? Menlo Park, Calif.: Risk Management Solutions.

Risk Management Solutions. 1995b. What if the 1906 Earthquake Strikes Again? A San Francisco Bay Scenario. Menlo Park, Calif.: Risk Management Solutions.

Robinson, Linda G., and Jack P. Gibson. 1989. Commercial Property Insurance. Dallas, Tex.: International Risk Management Institute.

Roth, Jr., Richard J. 1995-1996. California Earthquake Zoning and Probable Maximum Loss Evaluation Program. Los Angeles: California Insurance Department.

Rothschild, Michael, and Joseph Stiglitz. 1976. Equilibrium in competitive insurance markets: The economics of markets with imperfect information. Quarterly Journal of Economics 90:629–650.

Roy, A.D. 1952. Safety-first and the holding of assets. Econometrica 20:431–449.

Russell, Thomas, and Dwight M. Jaffee. 1995. Catastrophe insurance, capital markets, and insurable risk. Journal of Risk and Insurance 64:205–230.

Russell, Thomas, and Dwight Jaffee. 1996. Sharing the risk: Northridge and the financial sector. Paper prepared for EERI/FEMA Conference, "Analyzing Economic Impacts and Recovery from Urban Earthquakes," Pasadena, Calif., October 10–11.

Sauter, Franz F. 1996. Redefining terms in the field of seismic safety and risk mitigation. Earthquake Spectra 12(2):315–326.

Scawthorn, Charles. 1995. Insurance estimation: Performance after the Northridge earthquake. Contingencies, Sept./Oct., pp. 26–31.

Scism, Leslie. 1996. Florida homeowners find insurance pricey if they find it at all. Wall Street Journal. July 12, p. 1.

Simmons, Malcolm M. 1988. CRS Report to Congress: The Evolving National Flood Insurance Program. Report 88-641-ENR. Washington, D.C.: Congressional Research Service.

Slovic, Paul, Howard Kunreuther, and Gilbert F. White. 1974. Decision processes, rationality and adjustment to natural hazards. In Natural Hazards: Local, National and Global, Gilbert F. White, ed. New York: Oxford University Press.

Slovic, Paul, Baruch Fischoff, Sarah Lichtenstein, B. Corrigan, and B. Combs. 1977. Preference for insuring against probable small losses: insurance implications. Journal of Risk and Insurance 44:237–258.

Spulber, Daniel F. 1989. Regulation and Markets. Cambridge, Mass.: MIT Press.

State Board of Administration of Florida. 1996. Florida Hurricane Catastrophic Fund Report of Activities as of June 3, 1996. Tallahassee: State Board.

State of Florida. 1993. Report of the Study Commission on Property Insurance and Reinsurance. Tallahassee: State of Florida.

Steinbrugge, K. V. 1982. Earthquakes, Volcanoes, and Tsunamis: An Anatomy of Hazards. New York: Skandia American Group.

Steinbrugge, Karl V., and S. T. Algermissen. 1990. Earthquake Losses to Single-Family Dwellings: California Experience. U.S. Geological Survey Bulletin 1939-A. Denver, Colo.: USGS.

Steinbrugge, Karl V., and Richard J. Roth, Jr. 1994. Dwelling and Mobile Home Monetary Losses Due to the 1989 Loma Prieta, California, Earthquake with an Emphasis on Loss Estimation. U.S. Geological Survey Bulletin 1939-B. Denver, Colo.: USGS.

Stone, James. 1973. A theory of capacity and the insurance of catastrophe risks: Part I and Part II. Journal of Risk and Insurance 40:231–243 (Part I) and 40:339–355 (Part II).

Suttmeier, Richard P. 1994. Risk in China: Comparative and historical perspectives on its social construction and management. Technological Forecasting and Social Change 46:103–124.

Taylor, Craig E., Howard Kunreuther, and Daniel Alesch. 1994. Earthquake Insurability and Small Business. Prepared for the National Science Foundation under Award No. BCS-9100090. Los Angeles: Dames and Moore.

Taylor, Craig, Ronald Eguchi, and S. E. Chang. 1998. Updating real-time earthquake loss estimates: Methods, problems, and insights. In Proceedings of the Sixth U.S.. National Conference on Earthquake Engineering. Oakland, Calif.: Earthquake Engineering Research Institute.

Tversky, A., and D. Kahneman. 1974. Judgment under uncertainty: Heuristics and biases. X. Science 185:1124–1131.

Tversky, A., and D. Kahneman. 1981. The framing of decisions and the psychology of choice. Science 211:253–258.

Tversky, A., S. Sattath, and P. Slovic. 1988. Contingent weighting in judgment and choice. Psychological Review 95:371–384.

U.S. Congress. 1995. Federal Disaster Assistance. Report of the Senate Task Force on Funding Disaster Relief. Washington, D.C.: U.S. Government Printing Office.

U.S. Department of Housing and Urban Development (HUD), Office of the Secretary. 1966. Insurance and Other Programs for Financial Assistance to Flood Victims. A Report from the Secretary of Housing and Urban Development to the President, as Required by the Southeast Hurricane Disaster Relief Act of 1965. Washington, D.C.: U.S. Government Printing Office.

U.S. General Accounting Office. 1979. Issues and Needed Improvements in State Regulation of the Insurance Industry. Washington, D.C.: Government Printing Office.

U.S. General Accounting Office. 1992. Coastal Barriers: Development Occurring Despite Prohibitions Against Federal Assistance. GAO/RECD-92-15.Washington, D.C.: GAO.

U.S. General Accounting Office. 1993. Flood Insurance: Information on Various Aspects of the National Flood Insurance Program. GAO/T-RECD-93-70. Testimony Before the Subcommittee on Housing and Urban Affairs, Committee on Banking, Housing and Urban Affairs, United States Senate. Washington, D.C.: GAO.

U.S. General Accounting Office. 1995. Thomas McCool Testimony Before the Subcommittee on Water Resources and the Environment, Committee on Transportation and Infrastructure, House of Representatives, December 5, 1995, GAO/T-GGD-96-41. Washington, D.C.: GAO.

U.S. Senate, Committee on Banking and Currency. 1966. Insurance and Other Programs for Financial Assistance to Flood Victims. Washington, D.C.: U.S.. Government Printing Office.

Valery, Nicholas. 1995. Earthquake Engineering: Fear of trembling. The Economist 335(7911):SS3–SS5.

Vowinkel, Patricia. 1995. Costly '95 hurricane season may herald new cycle. Journal of Commerce, October 20.

Walker, George. 1996. Earthquake Insurance Issues. Paper presented at the 1996 Annual Conference of New Zealand National Society for Earthquake Engineering, New Plymouth, New Zealand.

Warfel, William J. 1994. Market failure in property insurance markets: The case for a joint region-federal disaster insurance program. Journal of Insurance Regulation 12:486–514.

Weinstein, Neil, ed. 1987. Taking Care: Understanding and Encouraging Self-Protective Behavior. New York: Cambridge University Press.

White, Gilbert. 1953. Human Adjustments to Floods. Research Paper No. 29. Chicago, Ill.: University of Chicago.

Wildavsky, Aaron, and Karl Dake. 1990. Theories of risk perception: Who fears what and why, Daedalus 119(4):41–60. Reprinted in Environmental Risks and Hazards, Susan Cutter, ed. Englewood Cliffs, N.J.: Prentice-Hall. 1994.

Will, Frederick L. 1974. Induction and Justification. Ithaca, N.Y.: Cornell University Press.

Winter, Ralph. 1988. The liability crisis and the dynamics of competitive insurance markets. Yale Journal of Regulation 5:455–500.

Winter, Ralph. 1991. The liability insurance market. Journal of Economic Perspectives 5:115–136.

Working Group on California Earthquake Probabilities. 1995. Seismic hazards in southern California: Probable earthquakes, 1994 to 2024. Bulletin of the Seismological Society of America 85(2):370–439.

Index